# WALK
## WITH
## DINOSAURS
# FASCINATING FACTS

# WALKING WITH DINOSAURS

## FASCINATING FACTS

### MIKE BENTON

London, New York, Sydney, Dehli, Paris,
Munich, and Johannesburg

Publisher: Sean Moore
Editorial director: LaVonne Carlson
Project editor: Barbara Minton
Editor: Jane Perlmutter
Art editor: Gus Yoo
Production director: David Proffit

This book is published to accompany the television series
*Walking with Dinosaurs* which was produced by the BBC
and first shown on BBC1 in 1999.

Executive producer: John Lynch
Series producer: Tim Haines
Producers: Tim Haines and Jasper James

First published 2000 by BBC Worldwide, Limited
Woodlands,
80 Wood Lane, London W12 0TT

ISBN: 0 7894-7168-X

Commissioning Editor: Joanne Osborn
Project Editor: Helena Caldon
Art Direction: Linda Blakemore
Book Design: Martin Hendry
Illustrations: John Sibbick

Set in Stone Sans
Printed and bound in Great Britain by Mackays of Chatham
Color separations by Radstock Reproductions Ltd, Midsomer Norton
Color sections printed by Lawrence Allen Ltd, Weston-super-Mare
Cover printed by Belmont Press Ltd, Northampton

# CONTENTS

# INTRODUCTION

*Walking with Dinosaurs* has created a huge amount of attention. This celebrated television series has now been seen by more than two hundred million people worldwide, and it is probably the most successful documentary series ever made for television.

Why the interest? The visual effects are spectacular – they present a world that is strange, marvelous, beautiful, but also real. This is not science fiction, but a serious attempt to show everyone just what life on our prehistoric earth really looked like.

*Walking with Dinosaurs* is a program created with adults in mind, but it has been watched and loved by both adult and junior dinosaur fanatics alike all around the globe. The images on the screen have raised huge numbers of questions from viewers since the program's first transmission, and the goal of this book is to answer them.

The accuracy of the *Walking with Dinosaurs* images sparked off a wide-ranging debate. Some paleontologists have complained that there are factual errors here and there (debates that will run on and on, no doubt), while others have complained about the whole project, suggesting that it was perhaps not a good idea even to attempt to show dinosaurs as living animals. After all, they argued, we only have the dry bones in the museums so how on earth can you bring them to life? Surely, they argued, it's all guesswork?

This has been a minority view, although one expressed vigorously in some quarters. Most geologists and paleontologists have been just as stunned as everyone else by the technical wizardry of the special effects; others have been more than happy to see this huge investment of money and expertise into their subject area. These scientists are aware that approximately 100 expert paleontologists from all parts of the world were consulted to check that details were correct. They are also aware that a great deal of cutting-edge, often not yet published, research went into the documentary series.

*Walking with Dinosaurs* did not simply walk a well-worn path by recycling old-fashioned images of dinosaurs. Some of the behavior and movements the dinosaurs and other prehistoric beasts exhibited in the television series have only recently been established. For example, as a result of first-hand experiments with spectacularly well-preserved skeletons and state-of-the-art bio-mechanical calculations, we know the pterosaurs walked awkwardly on all fours. These techniques were also applied to all the locomotion sequences; in many cases, the computer animators worked hand-in-hand with biomechanics and computing experts for weeks to get it right.

But in the end, how much of what was shown on the screen is real and how much is guesswork? In *Walking with Dinosaurs: Fascinating Facts,* you can read all the background details and discover how dinosaurs moved and fed, how they mated and looked after their young, how they fought and defended themselves, whether they were warm-blooded or not, what noises they might have made, and the latest thinking on why they died out.

This book will take you right into the field with the paleontologists as they dig up bones around the world and sort them out back in the lab. The text also explains the brilliant theoretical insights paleontologists had in order to find out the impossible. To the question, "How on earth can you make the bones come to life?" The answer is, "Mix together a great deal of anatomy, biomechanics, biology, chemistry, physics, and geology with a huge amount of wild inspiration, and then it's really quite easy."

A special feature of the book is the appendix, listing all the dinosaurs ever named. It's quite surprising that twenty or more new dinosaurs are named every year, and the sum total is now about 800. This is a reference resource that has never before been available, but its presentation here will be invaluable for the serious dinosaur enthusiast.

Michael J. Benton
May 2000

**CHAPTER ONE**

# FLESH ON THE BONES

It's amazing to think that the images in *Walking with Dinosaurs* are based only on fossils. Dinosaur fossils are ancient bones and pieces of bones that have been buried and infiltrated with minerals, and that is all. How on earth can paleontologists, the scientists who study fossils, bring the dry bones back to life? Are all these models, pictures, and moving images of dinosaurs just wild guesswork, as some critics have said, or is there a set of rules and insights that scientists use to breathe life into these ancient beasts?

## ● What is a fossil?

The word "fossil," from the Latin, means literally "dug up." (The root is used in "fosse," an archaeological term for a ditch.) At one time scientists used the word "fossil" to describe anything they dug up out of the ground – old coins, arrowheads, true fossils, even potatoes. Nowadays, the word "fossil" is restricted to anything that is the remains of a plant or animal that lived a long time ago.

## ● Digging up the bones

Dinosaur bones that you see in a museum have been hacked and chiseled from rock. At any one time there are probably ten or twenty dinosaur excavations going on somewhere in the world. They normally happen in summer, the traditional "field season," when professors and curators jump into their four-wheel-drive vehicles and set off into the wilds. The digs may be near to base, or they may be further afield. Recently, paleontologists have been setting off into the wastes of Africa, Australia, Mongolia, and even Antarctica, and that takes some organizing, and some money!

# Anyone can dig up dinosaurs

Most dinosaur diggers are students. Eager students, who may be studying geology or biology at a university, or any other subject, often volunteer to go along on a dig as laborers. But nowadays there are open digs in many countries, some of them organized on behalf of companies like Earthwatch, others simply by a university or a museum, where the paleontologist in charge has agreed to take on a number of untrained volunteers. The volunteers pay their basic costs and, in exchange, they have the chance to handle real dinosaur bones in the rocks, and to take part in a proper scientific dig. There's more to it than just swinging a pick and hoping for the best.

# Digging up a dinosaur costs a lot

It might cost any amount from $1600 to $800,000 (£1000 to £500,000) to dig up a dinosaur skeleton. At the cheaper end of the scale are sites that are close to base, where travel and living costs are inexpensive, and where the diggers perhaps pay their own way. Also, it helps if the rock is soft, if the days are long and sunny, and if the dinosaur diggers are young and active. At the top end of the scale would be a dinosaur excavation in Antarctica or Greenland, for example; traveling to such countries costs a lot, you would need to rent helicopters when you get there – and they cost a huge amount – and living costs might be high (extra survival equipment, fancy tents, and so on). Also, the rocks might be hard and difficult to remove.

# Who pays to dig up dinosaurs?

No single foundation has pots of money to dole out for dinosaur digs. The National Geographic Society is one of the best sponsors, commonly paying $10,000 to $50,000 towards the costs of an expedition. Sometimes, paleontologists can secure funds from a government-sponsored research fund or from some charitable foundation. At other times, the dig is essentially self-funded, out of money from normal university or museum budgets and from the diggers' own pockets.

# Getting to the dig site

Before the dinosaur-digging team sets off, its members need some basic equipment. Transportation is essential, usually one or more jeeps. You have to be able to get to the site, to drive around exploring for new sites, and get out again with the loot. Dinosaurs are rarely conveniently located beside major roads. Next, the team requires tents and other basic camping equipment. For a long distance trip, say to Mongolia or central Africa, the domestic supplies may have to include food, water, and other basics necessary for a long trip, and to cover emergencies. Paleontologists can die in pursuit of those bones!

## ● Dinosaur-digging kit

The well-equipped paleontologist, setting off in the field, requires hammers, chisels, crowbars, scrapers, mounted needles, brushes, collecting bags, drawing materials, cameras (still, movie), tables (for laying out specimens), and maybe a microscope or two. Increasingly, crews also go equipped with power tools (pneumatic drills, rock saws), and Global Positioning System (GPS) equipment (for locating and mapping sites by taking a precise fix from satellites).

## ● Finding where to dig

Having loaded your jeep and ready for the dig, what next? Paleontologists do not drive around at random. They will only be able to raise funding if they have a sure-fire locality. They rely on older accounts of where people once found bones in the past. Perhaps they spend time in advance looking for likely sites. Usually, nowadays, there is a prior history, even if it is just a report that a quarryman or a child found a scrap of bone decades ago. Clearly, dinosaurs are found only in rocks that were deposited during their tenure on Earth.

## ● Continental sedimentary rocks

Paleontologists focus on continental sedimentary rocks of Mesozoic age. Now there's a mouthful. "Continental" rocks are those that were laid down on land, in rivers or in ponds, but not in the sea (those would be marine rocks). Dinosaurs lived on land, so their bones will be found in ancient soils and in river and lake sediments, but only very, rarely will they be washed into the sea. "Sedimentary" rocks are rocks that were once sediment, in other words hardened these rocks were once muds, silts, sands, or lime muds. They are called mudstones, siltstones, sandstones, or limestones accordingly.

## ● Geological time

"Mesozoic," means "middle life," and it refers to the fact that the Mesozoic is the middle of three great divisions of geological time (see

| EON | ERA | PERIOD | EPOCH | DATE at beginning (Myr) |
|-----|-----|--------|-------|-------------------------|
| **Phanerozoic Eon** | | | | |
| | **Cenozoic Era** | | | |
| | | Quaternary Period | | |
| | | | Holocene Epoch | 0.01 |
| | | | Pleistocene Epoch | 1.6 |
| | | Tertiary Period | | |
| | | | Pliocene Epoch | 5 |
| | | | Miocene Epoch | 23 |
| | | | Oligocene Epoch | 35 |
| | | | Eocene Epoch | 56 |
| | | | Paleocene Epoch | 65 |
| | **Mesozoic Era (Age of Dinosaurs)** | | | |
| | | Cretaceous Period | | 146 |
| | | Jurassic Period | | 205 |
| | | Triassic Period | | 250 |
| | **Paleozoic Era** | | | |
| | | Permian Period | | 290 |
| | | Carboniferous Period | | 362 |
| | | Devonian Period | | 408 |
| | | Silurian Period | | 510 |
| | | Cambrian Period | | 550 |
| **Precambrian Eon** | | | | 4560 |

**The geological time scale, showing major divisions of Earth history. The eons are divided into eras. We are concerned mainly with the Mesozoic Era and the Triassic, Jurassic and Cretaceous periods.**

above): Paleozoic is "ancient life," and Cenozoic is "recent life." The Mesozoic era lasted from about 250 to 65 million years ago (geologists do not count backwards from the present day, but forwards from more ancient to more modern). Dinosaurs first evolved about 230 million years ago, in the middle of the Triassic Period, and lived in abundance through the Jurassic and Cretaceous periods, until they died out at the end of the Cretaceous, 65 million years ago (see Chapter 10).

# ● Prospecting

Once on site, some of the team will set off prospecting. They will walk for hours, up and down gullies, eyes glued to the ground. This can be a hugely absorbing activity, and you often find that hours have passed and you are miles from base when the hunger pangs strike. The point is that in good dinosaur country, there may be fragments and shards of bone all over the place, washed down by streams from their original residence spot far away. These fragments may be interesting in themselves – after all, a complete *Tyrannosaurus rex* tooth is certainly worth picking up.

## ● Finding the bones

For the dinosaur prospector, it is what the bone shards may hint at that is of great importance. Following bone scraps back up a gully may lead to a complete skeleton buried in sediment. The prospector marks the spot, and will try to assess how much of the skeleton may still be buried: has it been virtually all eroded away, or is there enough still there to make it worth while setting up a major dig on the site?

## ● Marking out the site

If the dinosaur prospect looks good – if, for example, there seems to be a line of bits of bone peeping out along a gully side, and extending for several yards, then there might be a complete skeleton there – the prospector might scrape away sediment from above the bone layer and try to identify some of the parts that are present. He or she will hammer in a post, perhaps take a precise GPS reading (which can define a ten-yard by ten-yard square) and go back to base. There's no point finding a great prospect and then not marking the location. Dinosaur country can be bleak and featureless.

## ● Removing the overburden

When a decision has been made to start a dinosaur dig at a particular spot, the biggest job is to remove the overburden. This is a great word, because it says it all. Typically, a skeleton will be peeping out of a sloping gully wall. The amount of overburden to be removed depends on the size of the buried skeleton, but also on the angle of the slope of the gully wall. If it is a steep slope, then a huge amount might have to be removed. The skeleton is likely to extend at least seven to ten feet into the gully wall, and that could mean removing 16 to 30 feet of overburden. The sediments above have to be removed thoroughly to make the site safe, and also to protect the bones as they are exposed. It is not helpful to leave precipitous piles of overburden above a site, where they could collapse back over and crush the specimen.

### ● Fossils in flat lands

Clearing the overburden from a typical dinosaur site might take a week or so of backbreaking work. Sometimes, however, collectors are lucky: for example, when they are working in the wide, flat deserts of the Gobi or the Sahara. There, the sediments are often soft, and the wind and moving dunes expose skeletons lying flat in front of you. Sometimes there is no overburden at all. But that is rare.

### ● Clearing the bones

Once the overburden has been safely removed, the paleontologists put away their shovels and power tools and go down on hands and knees with scrapers and brushes. If they have done their work well, they will have planned to stop a few inches above the bones themselves. Then, the painstaking handwork begins. Rock is chipped away carefully. The technique is to find where a bone is and to place the chisel off to the side before tapping it gently with a hammer. Then, if the chisel slips, you don't destroy a bone.

### ● Fine-detail site clearance

When paleontologists are working on a dinosaur site, sometimes a needle mounted in a wooden handle is the best implement. By careful probing and scraping, the sediment can be loosened, and it is then brushed away. As the bones are completely exposed from above, the collectors paint them with dilute preservative solutions, usually some kind of synthetic glue, dissolved in acetone, which soaks deep into the pores of the bone to harden it up. Doing all this work out in the open might seem rather foolhardy. Why not simply parcel up the rock and bones and take them back to the lab for cleaning there? This wouldn't work. First, you need to see where the bones are, before you can lift them, and you have to be able to record what's where on the spot.

### ● Mapping the site

Once the whole dinosaur skeleton is exposed, a key operation is to map it. The site is gridded, usually using thin strings, into one-yard squares. A quadrat is then used to subdivide the squares. The quadrat is made from one-yard-long laths of wood nailed into a perfect square. Some tough transparent plastic is stretched over the laths, then pinned down. Then a four-inch grid is drawn over the plastic, so the one-yard quadrat is divided into a hundred (ten by ten) squares. The paleontologists work in teams to map the site. They transfer the information from the ground onto to squared paper, perhaps scaling the one-yard squares to a four-inch square on the paper. Then the four-inch sub-squares, now on the plastic sheet, become 1.5-inch squares on the site drawing.

## ● Photographic records

While mapping and recording the dinosaur site, other crew members will be photographing everything from all angles, and recording the new find on video film as well. But the hand drawn map is essential. Photographs can rarely be taken of the whole site from directly above (I've never heard of a case of a paleontologist renting a balloon, but that would be the only way). Also, if the bones are the same color as the rock (and they often are), then photographs will not be much use when you get them home. The map is essential for piecing the skeleton together back in the lab, and also for interpreting how it was buried. Have bits of the skeleton, for example, been broken off and washed away? Is the skeleton lying in a curled-up posture, or is there anything else unusual which might show how it died?

## ● Getting the bones out

When dinosaur collectors first went into the American Midwest, they simply hacked the bones out with pickaxes and loaded them into horse-drawn carts like piles of logs. After all, the collectors were often laborers who were employed to build the trans-American railroads, and that's how they dealt with rocks. But the paleontologists back east were not very happy when they looked into the packing crates and found piles of dust and shattered debris. Then, in the 1880s, some of the field men figured out that if you set a fossil bone in plaster, it would be protected in just the same way as a broken leg set in plaster in a hospital. And their technique is still used today. Dinosaur bones may be big and heavy, but they can shatter like a piece of porcelain at the slightest jar and jolt.

## ● Getting the bones out of the ground

Once the dinosaur skeleton has been mapped, the first job in removing the bones is to survey how best to divide up the skeleton. The crew usually don't plaster up the whole thing in one block, because that would weigh tons. Normally, because of the way the bones lie, it is possible to draw lines between neighboring ones, and to divide the skeleton up into a number of separate parcels. Usually, too, you don't wrap each and every bone separately. In a typical dinosaur skeleton there are 300 bones, and that would be too time consuming.

## ● Bones on pedestals: first step to removing the bones

Once the division of the skeleton in the ground has been made, the members of the crew dig trenches around the marked blocks. They dig deep, maybe a yard or more, so that the plaster cast can be made properly. This process is called pedestaling, or putting the bones on pedestals. Then sheets of wet paper are placed over the bones to act as a separator before the plaster is applied. Paper towel is fine to use, too.

## ● Getting plastered

To protect the dinosaur bones before they are removed, a mixture of plaster of Paris is made up. Usually the crew will have bought several large sacks from a builders' supply. They mix up quarts of the stuff until it is sloppy. Then they run strips of sackcloth (burlap) through the gooey mixture and slap them over the bone in a crisscross pattern until there are maybe five to ten layers. The bigger the bone, the more layers are needed. After the plaster has dried (a few hours later), the bone in the rock is undermined and carefully flipped over. As much sediment as possible is prised out from the back of the plaster cast, and the open side (the former underside) is plastered over. When the parcel is complete, it can be carried out and stacked in the trucks.

## ● Moving huge blocks

Many dinosaur blocks are too heavy to carry. What do you do then? If the truck has some lifting gear, and you can get right up to the site, then that may be fine. But usually you can't do this, generally because the skeleton is in a very craggy spot which the vehicle cannot approach. Then the paleontologists have to do a bit of designing. One solution is that a sledge is made from beams of wood and then placed on the top of the partially plastered bone block. Then the sledge itself is plastered in place and left to dry. Once this plaster is dry, the whole block is flipped over with the sledge conveniently placed underneath.

## ● On the road gang

If a plastered block containing dinosaur bones is huge, then some civil engineering skills are required. On one dig in Alberta, we had to build a level road by hand down a steep slope to the side of a river, and it took ten people to haul the block out. On other digs, the crew builds trolleys and other strange wheeled contraptions to help them move vast weights. Remember that a cubic yard of rock weighs one to two tons. And a big dinosaur might be enclosed in a hundred or more cubic yards of rock.

## ● Trucking the bones home

Once the blocks containing the bones are on the truck, they can be taken home. In a long digging season, bones may be trucked out and stored at a base, or even taken all the way home to the laboratory. Even though they are protected by plaster casts, it is still wise not to toss the blocks about or subject them to any more jolting than is necessary.

## ● Cutting the plaster cast

Back home, after the dig, the plastered-up bones await treatment in the basement. Usually, the team members are eager to get the bones out and get at them. But, in practice, sometimes there are so many that it might take years before they can all be dealt with. The big museums around the world often have basements full of unopened plaster jackets (that's what paleontologists call the parcels) from a hundred years ago or more. Each plaster jacket has to be opened carefully, ideally with a dental cutting wheel, so that minimal damage is done. Initial work may be relatively rapid, since the worst of the plaster jacket and the surrounding rock (called, technically, the matrix) is chipped off. It's no surprise that well-equipped paleontology laboratories have massive suction and filtering systems to remove all the dust generated.

## ● Cleaning the bones

Once the worst of the plaster, rock and dust has been removed, the technician can settle to the immensely satisfying task of slowly cleaning and strengthening the bones. As rock is chipped away, the surface of the bone is exposed, and it may react badly to the air. In all cases, even though the bone has been filled with hard minerals, further dilute strengthening glues are soaked in to strengthen the bone. Sometimes a vacuum chamber has to be used to make sure that the strengthening liquid goes right through to the core.

## ● Reliefing

When a skeleton is small, it may not have all the rock removed from it. It is safer in these cases to expose it from one side and make it into a kind of relief sculpture, where it is backed by the sediment, which is cut back and smoothed, but the skeleton is then safely held in place. A new technique for preparation is the airbrasive machine. This is a high-pressure pump that blasts air out of a thin nozzle, like a pen, at very high pressure. Small beads of glass, or another substance, may be inserted in the machine. By moving the airpen over a delicate fossil, the matrix is virtually blown off before your eyes. No risk here of damaging anything, but it is slow, and only really useful on small specimens.

## ● Acid treatment

Sometimes the rock encasing a dinosaur skeleton is so tough that normal drills and chisels make no impact on it. The rock may contain iron minerals, which may be virtually impossible to clear, or the rock is hard because it is full of calcite (calcium carbonate), the basic component of limestone. In the 1950s, paleontologists at the Natural History Museum in London discovered that acid can eat away the calcium carbonate. So a bone in hard limestone can be placed in an acid bath, and a day or two later it will be completely clean, since the calcite will disolve into the acid. The technicians soon discovered that they had to be careful. If they tossed the rock and bone into any old acid, the whole block, bone included, might disappear by morning. This is because bone is made from calcium phosphate, which is also susceptible to solution in acid, although more slowly than calcite.

## ● Take care!

The normal practice now in removing bones from limestone is to use dilute acid, usually acetic acid (vinegar) or formic acid (the acid produced by ants, although it is made in chemical factories). The more famous acids – nitric, sulphuric, and hydrochloric – are too aggressive, and they are avoided. Acetic or formic acid at five percent dilution is about right. The technician thereby slows the solution process and can check progress daily, sometimes removing the bone, neutralizing it, strengthening it with glues, and then replacing it in the acid bath. The benefit of acid treatment is that the end result can be beautiful and it does not occupy a technician full-time on a single specimen.

## ● Identifying the bones

As the dinosaur bones are prepared in the laboratory, either
mechanically (i.e., using chisels, needles, and brushes) or chemically (by
acid), they are laid out for the paleontologist to inspect. It is hard to
stand back and wait until everything is ready, so the paleontologist, and
the students, will crowd round excitedly, discussing all kinds of theories
about what each bone is, what it comes from and so on. But their
theories are not wild guesswork. If they're well-trained in anatomy, they
can identify the bones right away. Dinosaurs had predictable skeletons,
just as humans and all other animals do.

## ● "The thighbone's connected to the knee bone"

Dinosaurs had a skull, a backbone, arms, and legs. The leg, for example,
had a single upper element, the thighbone, two shin bones, a set of
ankle bones, and a set of toe bones. After a few years of training, any
paleontology student worth his salt can tell at once if he is looking at a
femur or a tibia, a metatarsal or a fibula, and he might even be able to
pin it down to a particular group. After all, the tibia of each different
dinosaur group has a very specific shape, and particular knobs and
bumps. So if the bones are reasonably undamaged, it isn't too difficult
to narrow down just what is lying there on the bench.

## ● Narrowing down the identification

How on earth can a paleontologist identify her new dinosaur skeleton?
Once she's decided by visual inspection that it's a theropod, or an
ankylosaur, or a ceratopsian, she may have to go back to her library.
Paleontologists are bibliophiles, squirrels who love old books and
monographs. Some of the giant papers published over a hundred years
ago are just as useful now as they were then. The illustrations of the
bones are beautifully engraved, and the detail is superb. A dinosaur
paleontologist has to have a major library of her own – perhaps a
hundred basic paleontology and dinosaur books, several dozen
monographs, and hundreds of papers. The paleontologist pulls down
her boxfile of papers about ankylosaurs, say, and runs through them.

## ● Monographs and polyglot paleontologists

All the research work on dinosaurs that has been done up to now is
contained in scientific papers and monographs (literally "single
writings"). Monographs are the major publications in which the

anatomy of a whole animal is documented, or maybe a whole tribe of related forms. The monograph has been written by a single paleontologist, and it may be a doctoral thesis or a major piece of work that took several years. There will be drawings of all the bones and detailed anatomical descriptions to explain how each species differs from each other. The monograph has probably been published by one of the great museums of the world (*Bulletin of the American Museum of Natural History, Mémoir du Muséum National d'Histoire Naturelle, Paris, Trudy Paleontologischeskogo Instituta, Moskva*) and of course it may be written in English, French, Russian, Chinese – who knows? The paleontologist has to struggle through, trying to follow the text as well as she can.

## ● Naming a dinosaur

Normally, the new skeleton is clearly a member of a species that has already been named. But, as the excitement mounts, the paleontologist may see that it is something new. The skull may be a different shape, there may be extra spines or knobs along the back, or the legs may be significantly longer than anything yet described. Then it is the privilege of the discoverer to name the beast. There is no august international committee that makes up the names: the finder gets to do that. It is normal to form the name using Latin and Greek, and there are two parts, the genus name (with a capital letter) and the species name (with a lower-case letter), as in *Homo sapiens* or *Tyrannosaurus rex*. Part of the fun of it is that the name means something – here, "wise man" and "king tyrant lizard." You are not allowed to choose a name that's already been used for something else, of course, so you try to check. But mistakes can be made, because over a million and a half plants and animals, living and fossil, have now been named, and there is as yet no single foolproof repository of names (this will come, no doubt, through the Internet, but no one is rushing to compile it – just imagine typing out 1.5 million names, and cross-references, and then having to add another 20,000 or so every year).

## ● Premature species classification

Sometimes, of course, paleontologists are a bit premature, and they name their skeleton as a new species without checking everything. There may be a skeleton tucked away in a small museum in England or China that's already been named and is just the same. Sooner or later, though, someone will spot this and put things right. Putting it right is called "synonymization," announcing that name B is actually for the same beast that is called by name A. Date priority is critical, and that's why we have to call *Brontosaurus Apatosaurus*. The name *Apatosaurus* was published two years before *Brontosaurus*, but they're one and the same beast, so there we go: sentiment and appropriateness of a name don't count, just the date of publication.

## ● The main bones in a dinosaur

A typical dinosaur had many more bones than a human does – about 300 in all. The extra bones are the forty to fifty or so in the tail. Otherwise, a dinosaur had a skull, a backbone, ribs, arms, and legs, just like a human or any other backboned animal (vertebrate). The vertebrates, as a group, include the fish and the tetrapods (literally "four-footers"). Tetrapods comprise amphibians (like modern frogs and salamanders and their ancestors), reptiles (modern lizards, snakes, crocodiles, turtles, and various fossil forms, including dinosaurs), birds and mammals. All vertebrates have a skull and backbone, and all tetrapods have arms and legs, and the other necessary bits and pieces to tie all these parts of the skeleton together. A typical dinosaur skeleton is shown below.

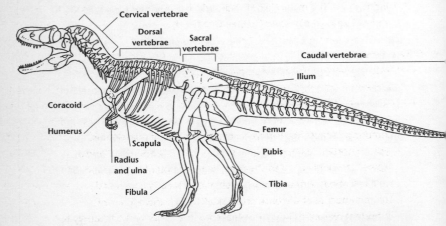

**The skeleton of the flesh-eating dinosaur *Allosaurus*, showing the major parts: the skull, the vertebral column (backbone), the limb girdles (shoulder girdle, pelvis), and limbs (arm, leg). All dinosaurs shared the same basic architecture.**

## ● The skull

The skull consists of two main parts: the outer protective part, which covers the nose region, the eyeballs, the jaw muscles, and so on, and the inner braincase. In humans and other mammals, the braincase has swollen out of all proportion to the head, and it is the main part of the skull, behind the face. But in reptiles the face is long and snoutlike, and the braincase (including the brain) are tiny. In fact, if the skull of a reptile is compared with a shoebox, the braincase is like a matchbox or cigarette packet shoved inside. And it's not very strongly attached, so in fossils the braincase sometimes gets washed away. In reptiles, the skull is made up of numerous thin platelike bones that fit together tightly along zigzag interlocking suture lines. The outer skull roof is tied together underneath by the palate, an arrangement of bony struts in the roof of the mouth.

## ● The lower jaw

The lower jaw of reptiles is made of five bones, whereas in mammals there is only one, the dentary. Reptiles also have the dentary, and that's the bone that carries the teeth. The other four reptilian bones form the inside and back region of the lower jaw and its hinging point with the skull. In dinosaurs, the lower jaws hinge right at the back of the skull, and they can move only like a hinge, which is important when trying to understand how dinosaurs fed (see Chapter 4).

## ● The backbone

As in humans, dinosaurs had a backbone, or vertebral column, made up of numerous separate bones, each one called a vertebra (plural, vertebrae). We use rather confusing words for these bones in common speech, calling the vertebral column the backbone or spine. These are confusing because the vertebral column is made of many bones, typically forty to a hundred, so it's more than a "backbone," and a "spine" is a single pointed structure, so more potential confusion. So we'll talk about vertebrae and vertebral columns. The vertebral column attaches at the back of the skull. In fact, there is a very important reason for this: the spinal cord, the major nervous supply to the whole body, runs out of the back of the brain and straight into a bony tube formed by the vertebral column. Apart from protecting the spinal cord, the vertebral column obviously supports the rest of the body. It's a major girder to which ribs attach, and the ribs form a supporting structure for all the internal organs, or guts, as well as helping to pump the lungs when the crisscross muscles that link the ribs make the whole ribcage swell up and out. The vertebral column is also a bridge to which the arms and legs are fixed.

## The tail

Humans don't have tails, however the remnant of the human tail is a small curl of tiny vertebrae tucked away inside your bottom: it's called the coccyx, and that's what hurts if you fall on your backside. Dinosaurs all had magnificent tails, usually made up of forty to fifty vertebrae. The tails had a variety of functions. In bipeds, the dinosaurs that walked on their hind legs, it was a balancing organ (see Chapter 5); in the big quadrupeds (that walked on all fours), the tail also acted for balance. Sometimes the tail had other functions: for example, whacking (see Chapter 8). In all dinosaurs, as in modern animals with tails, various important muscles for the hind legs attached along it.

## The shoulder girdle

The shoulder girdle is a fairly feeble structure in many ways. Humans are always breaking their collarbones (the narrow rod at the front at the top of your chest). At the back is the shoulder blade, or scapula. Dinosaurs had a shoulder blade that was similar in some ways, being attached to the ribs by muscles and tendons. The dinosaur scapula wrapped round the top half of the chest region, and a broad, platelike bone, the coracoid, ran round to the midline of the chest underneath. Other, thinner bones, including clavicles (our collarbones), were sometimes present in front. The key function of the shoulder girdle is to bind the arm joint firmly to the body.

## The arms

Our arms, like our legs, consist of a single upper bone, two lower bones, a bunch of small elements at the wrist and five long thin fingers made from separate joints. The upper arm bone (humerus) fits into the socket of the shoulder girdle. We can wave our arm around in all directions, an evolutionary hangover of our tree-dangling days. In four-footed dinosaurs, as in dogs and horses, the arm functions as a leg, but in bipedal dinosaurs, their arms could be quite flexible. The humerus attaches to the two forearm bones, the ulna and radius, at the elbow joint. The wrist in dinosaurs, as in humans, is made from five or ten cubic bones that give various kinds of flexibility. Primitive dinosaurs had five fingers, but later ones went down to four, three, or even two as was the case for *Tyrannosaurus rex*.

## ● The hip girdle

The hip girdle means business. When it is compared with the rather loosely attached shoulder girdle, the hip girdle is seen to be firmly fused to the backbone. The hip girdle of dinosaurs is made from three bones: the ilium at the top, the ischium below and pointing back, and the pubis below, and pointing back or forwards, depending on the group of dinosaur. (See figure, page 20.) The ilium is firmly cemented to specialized vertebrae, the so-called sacral vertebrae. Reptiles typically have two sacral vertebrae, but dinosaurs pretty well all had three or more, and some had as many as five or seven. So their hip girdles were *really* firmly fused to the backbone.

## ● The legs

Dinosaur legs are more like those of a horse or a dog than those of a human. There are essentially three crank elements, instead of our mere two. Let me explain. The upper crank, or lever segment, is the thighbone, the femur, which we have. The femur fits into the socket of the hip girdle and joins at the knee to the two bones of the calf, the tibia and fibula. In humans, that's it really, with the calf attaching via the ankle to the flat-down-on-the-ground foot. Dinosaurs, like horses and dogs, had an up-on-the-toes style of foot. In scientific terminology, we (and bears) have a plantigrade foot, while dinosaurs (and horses, dogs and cats) have a digitigrade foot. So these running forms all have an extra crank element in the leg, made from the bunched metatarsals. The dinosaur ankle is up off the ground, a kind of backward-pointing "knee." More of that later (see Chapter 5).

## ● Putting the skeleton together

Dinosaur students know all of this anatomy, and more. So when they are faced with a pile of loose bones, they can begin to put them together. If it's all from one skeleton, the ends of the bones will fit precisely: the femur of the left leg will articulate nicely with the left tibia, and you can manipulate the defunct dinosaur's knee joint. So, if everything is there, the complex three-dimensional jigsaw of the skeleton can be put together. If bits are missing, it may be possible to

sculpt them. For example, you need have only the left leg to know
precisely what the right leg looked like: a simple mirror image.

## ● Armatures

In classic skeleton mounts in museums, the real bones are strung
together. But being heavy, they have to be held up by a specially made
metal framework, or armature. This is a really skilled job, first fitting the
bones, assessing their natural postural angle, then firing up a blowtorch
and bending and welding strips of metal so they are the right shape,
but also strong enough to support the bones and slender enough to
be largely hidden when the skeleton is on show. This technique is not
so often used today, partly because of the complexity and cost, and
partly because of the risk of damaging the bones. A simple alternative
is to use casts.

## ● Casting the bones

It may seem like cheating to make casts from the original bones. When
you go to a museum, you like to know that the bones are real, that they
aren't just plastic models taken from other museums. Well, most
museums worth their salt will have collections of their own dinosaurs,
and they should display individual bones, some even for you to touch.
However, to make an effective mounted skeleton, lightweight casts
made of synthetic materials like fiberglass can be ten times better than
the real thing. The skeleton can be mounted in dynamic poses, and all
that's needed are hidden wires inside the bone casts. Indeed,
incomplete skeletons can be made whole again by duplicating missing
ribs or vertebrae, or by sculpting some elements. Also, a single skeleton
can become many. Some of the most dramatic modern displays may
have a whole herd of dinosaurs running side by side or attacking a
larger prey animal. Also, casts are useful for exchange.

## ● Comparative anatomy

But mounting a skeleton, or a cast of a skeleton, for a new display in a
museum isn't the end of it. In fact, it's really only the beginning. The
paleontologist urgently wants to know what the new dinosaur looks
like, how it lived, what it could do. To achieve this, the principles of
comparative anatomy come into play. Comparative anatomy is a grand
term for the rules of the structure of animals. We've already seen that
dinosaur skeletons aren't that different from ours. The human arm is in

some ways equivalent to the dinosaur arm. If a dinosaur has a sickle-like claw, it wasn't for picking its nose or for scratching its back. The paleontologist looks at modern animals with vast sickle-like claws and comes up, say, with lions and tigers on the one hand and sloths and anteaters on the other. So what was the dinosaur's claw used for – slashing flesh or for digging up ants? That requires further detailed comparative anatomical study of the shape of the claw, its cross sectional shape, its likely musculature and so on.

CHAPTER TWO

# IMAGES AND FASHION

It may seem odd, but the history of dinosaur images gives us a history of fashion – scientific fashion, that is. Paleontologists do their best when they advise an artist, but their knowledge is limited: the way a dinosaur looks, its pose, what it is doing, all depend on the current hypotheses. When you look back through old dinosaur books, you can pretty well date them by the dinosaur images. There was the giant lizard phase, the rhinoceros phase, the erect kangaroo phase, and the leaping biped phase of Victorian times. Perhaps, in a hundred years from now, people will look back indulgently at *Walking with Dinosaurs*, and say, "Well, they did their best, but it's quite ludicrous really." One hopes not. Certainly, by scanning back through old books and papers, it seems clear that views have changed, but that the swings of opinion are becoming less and less dramatic. Maybe we are homing in on the truth?

## ● Artistic reconstruction

There is a long and honorable tradition of dinosaur art, or paleo-art as it's sometimes called. At least with a skeleton, and with the muscles in place (see Chapter 1), and some careful thought about horns, spines and crests, skin patterns and colors, the paleontologist can collaborate with an artist to produce a painting of a life scene. Such images are prepared for museum displays, for books, for websites, even for technical descriptions. They combine a broad range of skills in the paleontologist and in the artist, and they may take some time to complete. But what an exciting objective. Having found your pile of bones, it is a wonderful end point to bring the dinosaur to life in a beautifully crafted painting!

## ● Dinosaur models

Paleontologists and artists have never wanted to stop at the two-dimensional image. Models of dinosaurs have been made since the 1850s, both full-sized models for museum displays and small pocket models for collectors young and old. Dinosaur models now must be some of the most common plastic toys around the world. Some of these are actually quite good, but...

## ● First dinosaur images

The first dinosaur images were produced in the 1830s, soon after the first dinosaurs had been named, in 1824 and 1825 (see Chapter 3). At that point the paleontologists Dean William Buckland of Oxford and Dr. Gideon Mantell of Lewes in Sussex had only isolated bones: bits, and pieces of two complete dinosaurs. It may come as a surprise to realize that these two were eager to promote their new fossil reptiles to the public. Indeed, Buckland, a senior academic and seemingly a crusty old chap, was a great artist and cartoonist himself. Paleontologists, from the start, have never been afraid to talk to the public and help people to understand the importance of the latest discoveries.

## ● Giant lizards?

What image of dinosaurs did the first discoverers have? William Buckland and Gideon Mantell thought they were simply huge lizards, each as long as five buses! The dinosaur teeth looked like lizard teeth, so these paleontologists thought that they had come from ancient giant lizards. By comparing the jaw bone of *Megalosaurus*, some eight inches long, with a modern lizard jaw, only a few millimeters long, Buckland simply scaled his lizard up. In a popular introduction to geology, the artist showed a weird lizard monster 200 feet long. Not even the wildest paleontological speculator today claims that size for any dinosaur!

## ● Rhinoceroses?

By 1840 the image had changed completely, from the nightmare lizard of the 1830s to an overblown rhinoceros. This was quite a shift, and it was not really based on any new evidence. By 1840 another four or five species of dinosaurs had been named, mainly from England, but also one from Germany. Richard Owen, a young man of the new generation, had the task of reviewing the state of paleontological knowledge up to that date. He thought about Buckland's and Mantell's Mesozoic

lizards and didn't like them. He rearranged the scattered bones of
*Megalosaurus* and *Iguanodon* and came up with a rhinoceros shape.
*Iguanodon* even had a horn for its nose. Or did it? But it would turn out
that Owen had his reasons, as we shall see.

## ● The Crystal Palace models

So successful was Owen that he was Sir Richard Owen by 1851, when
the Great Exhibition was held in London. This huge show captured the
public's imagination and showed them the wonders of science and
technology. A vast glass and steel structure had been built in the center
of London, in Hyde Park, the Crystal Palace, and everyone wanted to
preserve it somehow and create a permanent cathedral of science.
People began to think about dinosaurs.

## ● Prince Albert's plan

Prince Albert, a German by birth, was a science nut. He discussed
his enthusiasms with Queen Victoria (their pillow talk was usually in
German), but she wasn't really interested. But Prince Albert could see
that Britain's wealth was founded on its scientific and technological
advances. This was the time of railroad building, steam engines, huge
factories, the expansion of the British Empire. His idea was to create
a science park in south London, to move the Crystal Palace there, and
then create a landscaped educational garden around it. He called up
Professor Owen, and a plan was hatched. They would build the first
prehistoric park, and Britain's heroic rhino-dinos would be on show
at full size.

## ● Building the dinosaurs in concrete form

Sir Richard Owen engaged the well-known artist and sculptor
Waterhouse Hawkins, who was to be paid from the profits of the Great
Exhibition. Hawkins was commissioned to build a dozen prehistoric
beasts – dinosaurs, mammoths, and other primeval marvels. He chose
the new technology of reinforced concrete, a real innovation. The
models were to be life-sized, so he had to design a structure that was
light but strong. Metal frameworks were built, bricks were used to
create some of the inner spaces, and concrete was then poured section
by section to build up the body walls.

Sir Richard Owen's rhino-dino (1854). Owen believed that dinosaurs, were much more advanced than living lizards and crocodiles, and might even have been warm-blooded.

## ● The celebrity dinner in a dinosaur

Waterhouse Hawkins had his dinosaur models ready by December 1853. He kept the top of *Iguanodon*, like a lid, on one side. He had even painted the beasts, and they looked startling and strangely beautiful. A table was installed inside the *Iguanodon*, and a great celebratory dinner was held on New Year's Eve 1853. Owen, of course, occupied the chair of honor. Prince Albert excused himself from this spectacle. Professor Forbes, a geologist who attended the dinner, declaimed a poem he had written, and the assembled professors joined in the chorus:

> "The jolly old beast
> Is not deceased;
> There's life in him again."

## ● Owen's cunning plan

Why did Sir Richard Owen see the dinosaurs as giant rhinoceros-like reptiles? At first, people thought it was a chance move, or that he had new evidence from the fossil bones. He did not. But he had a very strong reason for promoting this rather advanced mammal-like image. Owen was essentially an anti-evolutionist. Progress was the buzz word, not only in British inventiveness but also in the fossil record: simple fossils came first, then more complex, and ever more complex as time went on. Owen hated that. He invented the rhino-dino to counter this progressionist view. Surely, he then said, reptiles have degenerated through time. Look at the dinosaur. It was a noble warm-blooded intelligent beast. Reptiles today creep on their bellies, they're slimy and unpleasant. We'll have no progress here! But Owen's vision was blown out of the water only four years later.

## ● Dr. Leidy's bipedal dinosaur

Attention shifted to North America in the late 1850s. Odd footprints and isolated bones and teeth had come to light in the United States along the east coast in the first half of the nineteenth century (see Chapter 3). But in 1858 Dr. Joseph Leidy of Philadelphia announced the first complete dinosaur. Remember, Owen had only isolated bones to deal with. Leidy obtained a skeleton of the duck-billed dinosaur *Hadrosaurus*, and he saw at once that it was a biped. It had long powerful hind-limbs, but the arms were short and had hands, not hooves. He had the skeleton set up in his new museum, and people could at last see what dinosaurs really looked like – or could they? Leidy got it right for the most part, but not quite.

## ● Tail-dragging erect kangaroos?

Joseph Leidy knew that *Hadrosaurus* was an erect biped – it stood up on its hind legs. It was pretty amazing to think that such huge creatures could actually balance somehow on their pins, and get about without falling over. Leidy cast around the modern world for a model to compare it with. *Hadrosaurus* might have stood like a human, a bird or a kangaroo. In the end, he opted for a cross between a human and a kangaroo. *Hadrosaurus* was mounted with its backbone nearly vertical, its head back, arms out at the front like a preying mantis, and legs straight down. The tail didn't quite fit, because it went straight down to the ground. What to do with the rest of the tail? It had to be effectively broken and bent at a right-angle to run horizontally along the floor.

## ● An American paleozoic park

The Americans were gaining confidence by the 1850s. They had their own scientists and their own industries. So they wanted their own Crystal Palace models too. Waterhouse Hawkins agreed to repeat his triumph, but he was to make all the dinosaurs in the new kangaroo pose. He sailed across the Atlantic and set up his studio in New York in 1868. The new "palaeozoic park" was to be set up in Central Park.

## ● "Boss" Tweed hated dinosaurs

Waterhouse Hawkins labored for three years building his models of the new American dinosaurs, and everything was ready, when his studio was visited by a gang of thugs who smashed everything in sight. The fragments of concrete were buried in Central Park. The thugs worked for William Marcy "Boss" Tweed, a notorious mobster who unofficially ruled New York. This was part of a power struggle with the local authorities, part an expression of "Boss" Tweed's anger about the primeval monsters that seemed to contradict his literal creationist view taken from the Bible. So ended the first American prehistoric park.

## ● Giant lizardlike monsters?

The debate about the dinosaur image rumbled on through the rest of the nineteenth century. One school of thought clung strongly to the old lizardlike image. They had complete skeletons now, so they couldn't show a modern lizard, simply scaled up to monster size. But they believed dinosaurs operated like lizards, somehow sluggish and creeping on their bellies. Models and pictures were produced in the 1890s and around 1900 showing the huge *Diplodocus,* belly to the ground, great legs bent at a bizarre angle out to the side. This extraordinary image had to bite the dust, though, because it was impossible: as the skeletons showed, you just could not bend the legs of *Diplodocus* in that way. And, as someone else noted, if it went along belly to the ground like that, it would have had to find a suitable trench two yards deep to accommodate its deep chest region. But the lizard supporters were already on the way out.

## ● Leaping *Laelaps*

By 1880 the bone-hunters Cope and Marsh had collected so many complete dinosaur skeletons from the American Midwest (see Chapter 3) that no one could go so far wrong in the future. A new image, more dynamic, began to come to the fore again. Paleontologists were particularly impressed by some of the small and medium-sized flesh-eating dinosaurs that were coming to light. Here were active little beasts with powerful legs, strong arms and beady little eyes. They had to be quick movers to catch their lunch. A famous image produced under the direction of Edward Cope shows two *Laelaps* (now synonymized with *Allosaurus*) leaping about in playful fashion. With this new image, we enter the twentieth century.

## ● The impact of Charles Knight

Dominant in creating a picture of dinosaurs that prevailed until the 1970s was Charles Knight. He got a job at the new American Museum of Natural History in the 1890s. He had a fine hand and a clear view, but no one at first knew how to use a paleo-artist. As they saw what the young Mr. Knight could do, his bosses became more and more enthusiastic. For the first time, they saw that a good museum could display the skeletons of dinosaurs, with great murals behind showing just what life had been like. Knight's paintings were accurate and realistic. His backgrounds were full of strange extinct trees and ferns. But his dinosaurs seem to leap out of the paper. They were vital and vigorous, not contorted into kangaroo or lizard postures. His pictures appeared in *National Geographic Magazine* as late as the 1940s, and they are still used in books today.

## ● Sinclair's dinosaur models

A new generation of 3-D models came in the early decades of the twentieth century. The oil company Sinclair adopted *Brontosaurus* as their icon in the 1920s. The company used dinosaur images in its advertisements; it handed out collector cards of dinosaurs and had dinosaur models at all its advertising shows. When the World's Fair was planned for Chicago in 1933, Sinclair paid for six full-sized dinosaur sculptures. They were made, again, from steel and concrete, and they included some of the biggest dinosaurs ever. For the first time, visitors could walk around a life-sized *Brontosaurus*. But the dinosaur action was not all American.

## ● Zdenek Burian

A Czech artist, Zdenek Burian, became fascinated by dinosaurs and ancient life as a child in the 1920s and 1930s. Even though his country had no dinosaurs of its own, Burian began a long and successful collaboration with the Czech paleontologist Zdenek Spinar. Together they produced a series of books in the 1940s to 1960s that were translated into many languages. His images appeared on stamps throughout the world, and they form an iconic set. Uniquely, Burian did not just paint pictures of dinosaurs, but his range spanned time from the origin of life to early humans.

## ● Dinosaurs in the movies

Dinosaurs didn't just stay on the printed page, in museums, or in the science fair; they quickly became staples of Hollywood. At first, film-makers didn't quite know how to work dinosaurs into their films, but by the 1930s a special kind of monster movie had been born.

## ● Gertie the first

The first dinosaur hero of the silver screen was Gertie, an amiable animated brontosaur. Gertie the dinosaur appeared in silent film days, in 1912, one of the first cartoons ever made. Dinosaurs, more famously, also appeared in Walt Disney's *Fantasia* in the 1940s, which showed a series of classic images of dinosaurs cavorting to the strains of classical music. And why not? These animated dinosaurs are light years away from the animations seen in *Walking with Dinosaurs*.

## ● Cross-dressing lizards

Film-makers wanted to put humans and dinosaurs together in the movies. How could that be done? (We know, of course, that humans and dinosaurs never lived together – dinosaurs died out sixty-five million years ago, and the first humans appeared about a million years ago, so there's a gap of sixty-four million years.) The first efforts involved trick photography and lizards and crocodiles wearing bits of cardboard. At least they move about, but the lizards and crocodiles with cardboard frills and horns in these films look just like that. Pretty unconvincing.

## ● The stop-action technique

After 1960 film-makers realized that no one was going to pay to see actors cavorting with overblown lizards with stick-on spikes. So they went to the other technique: stop-action filming. If you make a high-quality plastic model, take a still shot, move the model a bit, take another shot, move it a bit more, and photograph it again, you can eventually build up a moving film of a dinosaur walking across the screen, say. This stop-action technique needs a superb model-maker, and a huge amount of patience. Each setup might take ten or twenty minutes to create, and only the leg or arm should move. The rest of the beast has to be absolutely static.

## ● Ray Harryhausen and Raquel Welch

Hollywood employed a master craftsman in Ray Harryhausen, and he animated the plastic for many dinosaur movies from the 1950s onwards. His triumph was *One Million Years BC*, released in 1966. There were dozens of dinosaurs, some reasonably convincing cavepersons and also, of course, Raquel Welch. No wonder people often think that humans and dinosaurs lived together!

## ● Godzilla: guys in rubber suits

Movie dinosaurs can be recreated with costumes. The famous *Godzilla* movies were made in Japan on minuscule budgets. But how can you show giant dinosaur-like monsters fighting with each other, and wrecking cities, for only a few million yen? Copying the success of the Hollywood film of *King Kong* in the 1930s, the Japanese film-makers realized that a guy in a suit was the cut-price option. But the dinosaur had to look strangely human so the guy could fit in the suit: so was born Godzilla.

## ● Pulp dinosaurs

Dinosaurs penetrated the popular media in the 1920s. A whole genre of cheap comic books poured off American presses. Among the supermen and other action heroes, were a steady stream of dinosaur stories. All of them stemmed from Arthur Conan Doyle's *Lost World* and Edgar Rice Burrough's Tarzan stories. Beautiful women and muscular men were inexplicably lost in impenetrable jungles in some Africa/South America/Southeast Asia-type location. They were menaced by dinosaurs, snatched from the ground by huge pterosaurs, they wrestled great sea monsters, and were even sometimes attacked by a stray mammoth that had somehow got into the scene.

## ● Dinosaur revolution

Media dinosaurs were given a huge boost by a major revolution in dinosaur paleobiology which began in about 1970. In the late 1960s a new, younger generation of paleontologists was coming to the fore. With the expansion of the universities around the world, classes of graduate students were picking over long-forgotten fields of dinosaur research and sometimes coming up with quite radical ideas. One, Bob Bakker, a student at Yale University, looked again at the small flesh-eating dinosaurs. He had a mental image of active warm-blooded animals, not the grotesque cold-blooded behemoths that had become fixed in the mind of paleobiologists and public alike. He made a drawing that became an icon.

## ● Bob Bakker's iconic *Deinonychus*

One of the most important dinosaur images of the late twentieth century is a pencil sketch by Bob Bakker of *Deinonychus*, a modest-sized flesh-eater that had been discovered by his boss, Professor John Ostrom,

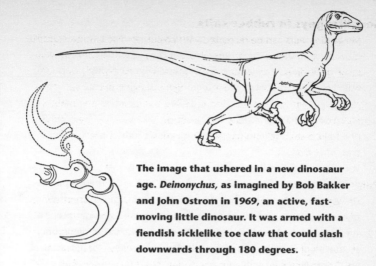

**The image that ushered in a new dinosaaur age.** *Deinonychus,* **as imagined by Bob Bakker and John Ostrom in 1969, an active, fast-moving little dinosaur. It was armed with a fiendish sicklelike toe claw that could slash downwards through 180 degrees.**

in 1964. *Deinonychus* was slender, and it had a huge flick-knife claw on its foot (see figure above). Bakker was full of ideas about dinosaurian paleobiology. He was also a superb artist. His image of the sleek, fast-moving *Deinonychus,* published as the frontispiece to Professor Ostrom's magisterial, but technical, monograph about the new dinosaur, set the scene for a new age.

## ● Dinosaurs as seesaws, not kangaroos

Bob Bakker realized in the late 1960s – and it seems so obvious that it's amazing it had been missed before – that a bipedal dinosaur has to be balanced over its hips. In other words, a bipedal dinosaur is like a seesaw, not a kangaroo. Leidy's upright kangaroo had long been abandoned, but Knight, Burian, and the film-makers always showed *Tyrannosaurus* stalking along with its backbone at forty-five degrees. Bakker saw that the backbone had to be absolutely horizontal. Then, like a seesaw, you could mark the pivot point right in the middle, over the hips. The front half had to equal the back half in weight. At first, the front half of a bipedal dinosaur looks bigger than the tail, but remember the cavernous lungs in the chest. They weigh nothing, so the balance is kept.

## ● Active leapers

The new predator *Deinonychus* was doubly interesting. Not only was it a sleek horizontal mover but it also had other capabilities. Ostrom and

Bakker speculated that the flick-knife foot claw was used for slashing prey. But the claw was on the foot. So *Deinonychus* had to be able to balance on one leg while it operated the claw. This meant warm blood, just like birds and mammals (well, maybe, maybe not: see Chapter 6). But at least here was a dinosaur that had to be able to leap and pivot.

## ● High-speed sauropods?

The sauropods, the long-necked, long-tailed monster plant-eaters, were ripe for re-evaluation in the early 1970s. If *Deinonychus* moved at speed, then so, too, did all the dinosaurs. In 1971 Bob Bakker published his arguments that the Mesozoic monsters, all 56 tons of them, could rear up on their hind legs to reach the highest, most succulent morsels in tall trees. He also imagined them galloping across the American plains at the speed of express trains. Where was this new world of dinosaurian paleobiology heading? It was heading for the buffers – or at least in part.

## ● The backlash in the 1970s

The 56 ton sauropod *Brachiosaurus* obviously could not have galloped. Its legs would have broken – simple as that (see Chapter 5). Paleobiologists, excited at first by the new dynamic dinosaur, realized that the enthusiasts had perhaps gone too far. There was a backlash, and a reconsideration of the biomechanical probability of each kind of model. But some of the new ideas seemed thoroughly reasonable.

## ● Head-crashing and fighting

The new generation of dinosaurian paleobiologists looked around at other monsters in the 1970s. They reanalyzed the armored dinosaurs. Some plant-eating dinosaurs had thickened skulls; others had horns and spikes on their heads. In days gone by, paleobiologists had simply said, "Oh, they're all for fighting." But what does that mean? By comparison with living mammals, it seems that some of these armored forms used their heads in tussling mock fights and as signaling devices to scare off other males from their patch, for example. Dinosaurian sociobiology had been born (see Chapter 8).

## ● Modern paleo-artists

Bob Bakker, founder of the revolution in dinosaurian paleobiology in the 1970s, was both paleontologist and paleo-artist. His new ideas and his enthusiasm were infectious. Knight and Burian represented a former generation, and new, young paleo-artists came on the scene, mainly in North America but also in other parts of the world. The new excitement in the science fed through to the public in a rash of books and exhibitions. The paleo-artists found commissions for their work, and hundreds of exciting and beautiful new dinosaur pictures were created.

## ● Moving dinosaurs

The excitement of the new dynamic dinosaurs also fed through to the model-makers. The first moving dinosaurs were exhibited in America in 1933, but these were rather crude, clunky affairs. They were built around heavy steel frames, covered with paper and wooden "skin." Their dinosour mouths opened and arms waved, but you could hear the engines inside clanking and whirring.

## ● Japanese wizardry

In the 1970s and 1980s, with advances in electronics and in miniaturization, it became possible to make convincing moving dinosaur models. Small motors are located here and there on a light metal frame, and the moving parts – heads, eyes, tongues, jaws, arms, legs – are attached to cranks and can be made to move in time to a soundtrack of roaring or whatever. The metal works and motors are clothed in a molded self-colored rubber "skin." These moving dinosaurs make marvelous museum exhibits, and several firms in Japan and the United States have specialized in constructing ever more elaborate examples.

## ● *Jurassic Park*

It had long been every film-maker's dream to create really convincing lifelike moving dinosaurs. New real-life animation techniques had been developed by about 1990, but they still had to be done using large mainframe computers. That made things difficult. But the first high-quality dinosaur animations were seen in Steven Spielberg's *Jurassic Park* in 1993. His animators worked long and hard and had to contend with incredible computing problems to pull it off. Things are somewhat easier now, but its still a pretty daunting job.

## ● Making *Walking with Dinosaurs*

In 1997 Tim Haines, a documentary producer at the BBC Science Unit, thought he would like to try to make a realistic animated documentary series about dinosaurs. Haines had trained as a biologist, and he saw this as the ultimate kind of documentary. He approached Mike Milne at Framestore, a computer animation studio in London that had made a reputation producing trick shots in television advertisements. The prospect seemed impossible.

## ● Raising the cash

In order to make *Walking with Dinosaurs*, it was obvious that there had to be two to three hours of computer animation, and for a minuscule price: only $9.6 million, approximately, (£6 million). (Steven Spielberg had spent more on only ten or fifteen minutes of animation in *Jurassic Park*.) But computer equipment and software had advanced a huge amount since the early 1990s. It might just be possible. The money was raised from a co-production group of British, American, German, and Japanese TV stations. But then the film-makers and computer animators had to speak to the paleontologists. Was it really possible to show living 3-D dinosaurs, complete with flesh and skin? Just how confident were the paleontologists?

## ● Putting the muscles in place

Paleontologists, perhaps surprisingly, are pretty confident about the fleshed-out body shapes of dinosaurs that you see in paintings and in *Walking with Dinosaurs*. Not only are skeletons pretty uniform through the tetrapods (see Chapter 1), but so, too, are the muscles. We have particular muscles to move the arms and legs, for example, in the arm, the biceps (the muscle that bodybuilders like to flex in front of you) is used for pulling the arm forward, the triceps, on the back of the arm, for pulling it back. Dinosaurs had similar muscles. The thickness of the bones gives an indication of the size of the muscles, and there are often roughened markings on fossil bones that show where muscles once attached. More on this in Chapter 5. Muscles are important: they are the flesh on an animal, and they provide the overall body shape. But what do we know about the skin?

## ● Dinosaur skin texture

There are actually some fossils that show the surface texture of dinosaur skin. Preserved perhaps by suddenly being overwhelmed by a sand-dune, one or two dinosaurs were mummified by drying out. As with the ancient Egyptian mummies, drying can preserve soft tissues for a long time without decay. The mummified dinosaurs did eventually lose their flesh, but only after the sand that had buried them had turned to sandstone and taken an impression of the dried skin. Dinosaurs all had largish scales in regular patterns in their skin. Some recent finds from China have shown that some flesh-eating dinosaurs even had feathers, or something like feathers on their bodies (see Chapter 9).

## ● Dinosaur skin color

Paleontologists know nothing about the skin color of dinosaurs. Color is almost never preserved (although there are plenty of fossil seashells, insects, and even turtle shells which show color patterns – lines of light and dark). So what about all those colors we see in all the reconstructions, including in *Walking with Dinosaurs*? Are they totally wild guesses, not really meaning anything at all? In a way, yes, but in a way, no. What can be said is that paleontologists assume that dinosaurs were colored in different ways for a reason. By studying modern animals, it's well known that color may be for camouflage (irregular blotches), for warning (bright stripes or flashes), for heat control (maybe dull matt colors) and so on. So paleontologists try to choose color patterns to match the environments and the postulated functions. But of course, for all we know, *Diplodocus* might have sported a delicious Gordon tartan, and *Allosaurus* might have been a delicate pink all over with purple and blue blotches.

## ● Scanning the model

The first step in creating living, moving dinosaurs is to build an accurate scale model in clay. This model may be only three feet or so in length, but every detail has to be sculpted precisely in clay or plastic. These models are called maquettes, using the French term for a model. The maquette is scanned using a laser scanner, which records every detail of the three-dimensional shape of the model inside a computer. This model is then stored.

## ● Making a stick model

The first step in the computer animation process is to turn the dinosaur into a stick model. This is like the stick figures children draw. It is a very simplified structure that reflects accurately the length of each element of the arms and legs, body, neck, tail, and head. The kinds of movements that each joint can make are then mapped on to the stick dinosaur. The stick model can be made to walk, jump, run, and swing around in an accurate way, reflecting current knowledge about exactly how that dinosaur operated (see Chapters 4 and 5).

## ● Making the model three-dimensional

Once a walking, jumping stick model has been created, it is then combined with information from the scanned maquette. The three-dimensional dinosaur model is dissected into curved segments, and these are fitted over the stick model. Now, when it moves, the segments flex and move, something like a crude suit of medieval armor. But the dinosaurs still look clunky, and they're just smooth and gray to the eye.

## ● Clothing the model with skin

For *Walking with Dinosaurs*, the skin stage was critical. If the skin lacked detail, the dinosaurs would look unrealistic. The skin had to have enough detail of texture and color so that it could be shown side by side with photographed modern scenes, and so that it could be shown in close-up. The film-makers employed a skin designer (nice work if you can get it!), who painted very detailed skin shapes for each dinosaur. His skin designs looked rather like the skin rugs old colonial explorers used to hang on their walls or lay on their floors. The skins designs were then scanned into the computers and wrapped around the gray dinosaur images and stuck firm. So when the dinosaur moved (driven by its hidden internal stick model), the whole beast now moved in a completely lifelike way. And while all this was going on?

## ● Backgrounds

BBC film crews spent a long time scouring the world for suitable backdrops. All the scenes you saw in *Walking with Dinosaurs* were real; only the dinosaurs were created by computer. The film crews were looking for modern locations where they could film prehistoric plants. They went to places like the Monkey Puzzle forest in southern Chile and the lush forests of Tasmania. They had to film the scenes as moving images, panning from right to left, swooping up into the trees, zooming in from a distance to a river bank. If the story line called for a dinosaur to run through a lake, someone had to create those splashes in the right places at the right times. If a dinosaur were to chew at a tree, someone had to shake the branch about in a realistic way.

## ● Special effects

The first efforts at combining animated dinosaurs and landscapes looked a bit wooden and unrealistic. The film of real landscapes was scanned into the computer, and the moving dinosaurs were put in place. The dinosaur movements were then refined and matched precisely to the backgrounds. If a dinosaur ran behind a tree, the effect had to be created by automatically masking part of the animated dinosaur as it ran past.

## ● Shadows

After the moving dinosaurs had been placed in their landscapes, shadow effects were created, using another computer program, so the dinosaur itself would move from light to dark in a dappled forest floor, for example, and so that each beast would throw a shadow on the ground. In places, the dinosaur was made to go out of focus, perhaps when it was running fast or when it moved out of shot. As a final stage, bits of plants were put in the foreground, so that there was some depth to the picture, a real background, then an illusory dinosaur on top, and then a real plant pasted on in front. This all sounds clunky, but when it's done by a skilful technician, the effects can be stunning.

## ● Other-worldly screeching

The last stage in creating *Walking with Dinosaurs* was to add a soundtrack. Again, who can say what kinds of noises a dinosaur made? (Although, surprisingly, there are some dinosaurs where we sort of know how they howled and bellowed – more of that in Chapter 8.) Music was written to reflect the action, and a commentary was added. Then a sound technician was given the whole film and told to make it sound good. He had the great job of filling up all the gaps with weird howls and growls. He had to add specific sounds if a dinosaur was in the scene, scruffling in the leaves, thumping as a heavy beast trotted past. But in the spaces between, he decided to make the landscape simply sound weird. When you watch *Walking with Dinosaurs* again, just listen for the strangled screeches and booming bellows as the camera pans across an ancient landscape.

CHAPTER THREE

# FINDING DINOSAURS

We've seen already how you dig up a dinosaur (Chapter 1). Well, those techniques have been pretty standard since the great American dinosaur rush of the 1870s and 1880s, except that we use four-wheel-drive vehicles now instead of horses. But when were the first dinosaurs found? Each new discovery affected the image of these creatures through the centuries, and we've already met some of the experts in Chapter 2. By the year 2000, over a thousand dinosaur species had been named (see Appendix). Who found them, and where did they come from? In this chapter we'll look at the history of dinosaur paleontology, starting from the earliest days.

## ● Classical Greeks

It is almost certain that giant bones were dug up around the Mediterranean. The ancient Greek philosophers millennia ago wrote about dragons and strange monsters. But these were probably the bones of large mammals. The Mediterranean, geologically speaking, is quite modern. It came after the Mesozoic, and the giant bones seen by the ancients came from mammals that lived ten or twenty million years ago. Aristotle and Plato probably never saw a dinosaur bone.

## ● Dragons' bones

Perhaps the first dinosaur bones were studied by ancient Chinese philosophers thousands of years ago. There may be a bit of dinosaur in the Chinese dragon. Dragons have been a part of Chinese culture for thousands of years, and who can say whether some Chinese dinosaur bones contributed to the stories? Certainly, it has long been the tradition to grind up fossil bones as a medicine in China.

## ● Plastic forces

In medieval times, philosophers debated some knotty problems. For example, quarrymen often dug out fossil seashells in limestone quarries that were high in the Italian hills. How had they got there? Were these stony petrifactions related to modern shells and fish, or were they simply odd pebbles that happened to look like the remains of plants and animals? A popular view was that fossils were "sports of nature" formed in the rocks by so-called "plastic forces." There are no such things as "plastic forces," but the grander philosophers talked about a *vis plastica*, hiding their ignorance in a bit of Latin mumbo-jumbo.

## ● The genius of Leonardo

As in so many things, Leonardo da Vinci got it right. In private notebooks that he wrote around 1500 he argued that the seashells on Italian mountaintops proved that the sea had deposited the limestones. They were not remnants of the lunch of Roman legionaries (another one of the ideas), nor had they been produced by plastic forces. Leonardo was ahead of his time, but the *vis plastica* theory was finally put to rest only in the eighteenth century.

## ● The first recorded dinosaur find

The first dinosaur bone to be described was found while the debate about the nature of fossils raged. Robert Plot, Professor of "Chymistry" at the University of Oxford, was preparing a monumental book on the "Natural history of Oxfordshire," and local naturalists sent him unusual specimens. These included a weighty rock that had been collected in a shallow limestone quarry at Cornwell in north Oxfordshire (see figure on p44). He included this specimen in an illustration of strangely shaped stones, some of which he interpreted as the preserved kidneys, hearts, and feet of humans. His interpretation of the rock from Cornwell was, however, quite different.

## ● The legs of giant men or women

Robert Plot, in trying to understand the huge rock, saw that the specimen looked like a bone. It had a broken end that was circular, and it seemed to have a hollow core that was full of sand. The unbroken end looked just like the end of a thighbone. Plot stated that it came from an animal that was larger than an ox or horse, and he considered the possibility that it might have come from an elephant brought to

**Robert Plot (right) illustrated the first dinosaur bone in his _Natural History of Oxfordshire_ (1676) – later called _Scrotum humanum_.**

Britain by the Romans. He ruled out that possibility since the bone was even bigger than that of an elephant. He had seen a living elephant at a traveling show in Oxford a year or two before. Plot's final decision was that the Cornwell bone came from a giant man or woman. He referred to mythical, historical, and Biblical authority in support of this interpretation ("There were giants in those days").

## ● _Scrotum humanum_, the first named dinosaur

Robert Plot's huge bone can be identified from his illustration as the lower end of the thighbone of _Megalosaurus_, a dinosaur that is now relatively well known from the Middle Jurassic of Oxfordshire. Plot's specimen is lost, but there is a twist in the tail. It was illustrated again in 1763 by R. Brookes, who named it _Scrotum humanum_ in honor of its appearance. Some giant! This is the first named dinosaur, although unfortunately the name has never been used seriously.

## ● The idea of extinction

Up until 1750, most naturalists believed that fossils represented species that were still living. If the fossil were quite unlike anything they knew, the naturalists thought that the species would soon be found in some unexplored region of the earth. After all, the naturalists all lived in Europe, and travelers were only then beginning to explore North and South America, Africa, and Asia. In a way, thought the naturalists, if we say some species is actually extinct, that would be a criticism of the wisdom of the Lord: He must have made a mistake. But huge bones began to be sent back to Europe from explorers in North America and Siberia. These began to shake people's faith.

## ● The American *incognitum*

The idea of extinction was gradually accepted after the discovery of North American elephants in about 1750. Explorers in North America began to dig up the remains of mastodons and mammoths, and sent some of the bones to Paris and London. There, distinguished anatomists and naturalists debated the specimens. Could these huge beasts still be living somewhere in the unexplored American West? Year by year, it became clear that they were not. Eventually, by 1800, nearly everyone accepted that extinction had happened.

## ● Archbishop Ussher's important calculation

How old was the earth? The Bible said that God created the world in seven days, but when did that important week take place? In the seventeenth century, Archbishop James Ussher of Armagh worked his way through the Bible, adding up the ages of X who begat Y, and Y who begat Z, all the way back from Jesus to Adam and Eve. He came up with the year 4004 BC. Some people still believe that this is the truth.

## ● The vastness of geological time

By 1750 naturalists were beginning to doubt the good Archbishop Ussher's calculation. They didn't doubt that he'd done his sums correctly, but they thought that perhaps the Bible was not really a scientific textbook. Perhaps some of the stories were more myths and allegories than hard factual accounts in every detail. Geologists who looked into quarries or scanned high sea cliffs wondered at the vast thicknesses of the rocks. They saw that many of these piles of rocks had accumulated slowly, like the mud and sand on the bed of the sea. They were looking at millions, not thousands, of years.

## ● Monsieur Cuvier's useful insight

Georges Cuvier (1769–1832) was a brilliant naturalist, and he is often called (certainly by the French) the father of paleontology, the father of geology, indeed the father of comparative anatomy. He established himself in a powerful political position in Paris during the 1789 French revolution, and also in Napoleon's times thereafter. Whether he could actually be father of all these sciences is debatable, but his greatest contribution was to point out the regularities of the structure of animals and give us the study of comparative anatomy.

## ● The idea of comparative anatomy

In about 1800, Georges Cuvier carried out a painstaking comparison of every bone of many fossil and living animals and noted similarities and differences between equivalent elements in their skeletons. He found that the shapes of bones indicated the purposes for which they were used and the relationships of the animals in question. By the 1820s, Cuvier had honed his skills in comparative anatomy to such a pitch of perfection that it was said he could identify any animal from any isolated fragment and that he could reconstruct any unknown fossil form from a single bone. In a famous public demonstration, he described a new fossil from quarries near Paris. Based on a few scraps, he declared, "Here we have in France a fossil marsupial." This was ridiculous since marsupials (opossums, koalas, kangaroos) are known today only in Australia and South America. But soon more complete skeletons of the small, pouched mammal were found, and the power of Cuvier's method was proved.

## ● Dean Buckland's eclectic diet

William Buckland (1784–1856) was both Professor of Geology at the University of Oxford and Dean of Christ Church. It was not uncommon for churchmen to combine their careers with science in those days. Buckland was interested in everything: he studied the geology around Oxford, he investigated caves full of mammal bones in Yorkshire, he was particularly intrigued by coprolites (fossil dung). He also had a broad diet: he ate any animal he could lay his hands on. Guests at his dinner table might be faced with cockroaches on toast to a start, boiled hamsters for the main course, and ants and cream for dessert. He admitted later in life that he really couldn't stomach flies.

## ● Buckland's bones

Dean Buckland was shown some collections of bones and teeth of a large fossil animal in about 1818. But he could not identify the bones, so he showed them to Georges Cuvier in Paris and to other experts. The best specimen was a jawbone with several long curved teeth. In the end, Buckland classified the animal as a giant reptile, probably, he thought, a lizard. After six years of consideration, Buckland finally published a description of the bones in 1824, and he stated that they came from a giant reptile that he named *Megalosaurus* ("big reptile"). This was the first dinosaur to be described.

## Mantell's Wealden rhinoceros

In the 1820s Gideon Mantell (1790–1852), a country physician in Sussex, was amassing large collections of Mesozoic fossils. During a visit to a patient near Cuckfield, so the story goes, his wife Mary, who had gone with him, picked up some large teeth from a pile of road-builders' rubble from the local Wealden rock formations. Mantell realized the teeth came from some large plant-eating animal, and, when he sent them to Cuvier, the great French anatomist assured him that the animal must have been a rhinoceros.

## Dinosaur number two

Gideon Mantell compared the fossil teeth he had found with those of modern animals in the Hunterian Museum in London, and a student there, Samuel Stutchbury, showed him that they were like the teeth of a modern plant-eating lizard, the iguana, only the fossil teeth were much bigger. Mantell described the second dinosaur, named by him *Iguanodon* ("iguana tooth") in 1825, based on the teeth and some other bones he had found since.

## Giant lizards

As we have seen in Chapter 2, Buckland and Mantell classified *Megalosaurus* and *Iguanodon* as giant lizards up to 200 feet in length. And why not? This was much more likely than Cuvier's first suggestion that, for example, *Iguanodon* was a rhinoceros. A rhinoceros in the Wealden? Everyone knew the Wealden were Cretaceous in age, i.e. Mesozoic. So they must be huge reptiles, lizards, or crocodiles, and they plumped for crocodiles.

## Dinosaurs three, four, and five

The third dinosaur to be named, *Hylaeosaurus*, came to light in southern England in the same Wealden beds that had produced *Iguanodon*. Gideon Mantell named it in 1833. Further dinosaurs were named in the 1830s, including two from the Triassic: *Thecodontosaurus* was named in 1836 by Henry Riley and Samuel Stutchbury from Bristol, England and *Plateosaurus* in 1837 by Hermann von Meyer from southern

Germany. Still, no one at that time knew just what these huge
Mesozoic beasts really were.

## ● Young Mr. Owen's financial support

Richard Owen (1804–92) was a confident and ambitious young man in
the 1830s. He became Professor of Anatomy at the Royal College of
Surgeons in London as a relatively young man, and he wanted a big
challenge. He was set the task of reviewing everything that was then
known about the giant reptiles of the land, sea, and air, and he was paid
$320, approximately, (£200) by the British Association for the
Advancement of Science to do so. Not much, but this was the first grant
of money given to any scientist in the field of paleontology!

## ● The Dinosauria are born

In about 1840 Richard Owen had some difficulty in understanding the
Mesozoic land reptiles: some were clearly crocodiles, others were small
and lizardlike, but there were all those giant "lizards." What was he to
do? At first, he tried to shoehorn them all into existing groups: some
were overblown lizards, others were crocodiles, and yet others were
something in between. Finally, he turned that idea on its head when, in
1842, he argued that some of them at least represented a new group,
which he called the Dinosauria (meaning "fearfully great reptiles"),
characterized by being terrestrial, being huge, and by having more than
two sacral vertebrae (see Chapter 1).

## ● A new image

As we saw in Chapter 2, Richard Owen had his own particular reasons
for "creating" his Dinosauria. Modern reptiles, had to be seen as a
degenerate remnant of this once-great clan, and Owen used this as
evidence against progressionism and evolution. We have also seen
how Owen's new image of dinosaurs dominated the Crystal Palace
park exhibition in 1854.

## ● Charles Darwin

Charles Darwin (1809–82) was not a paleontologist, yet his work on
evolution revolutionized the way in which paleontologists viewed their
fossils. Darwin began with a clear view that life had been created, as
recounted in the Bible. But on his voyage round the world he saw that
animals and plants that were on islands close to each other were

similar, and that fossils were often closely related to living forms in the same area. This suggested lines of relationship. Fossils were not just isolated objects, but instead they represented parts of a single great evolutionary tree that linked all species together from the origin of life to the present day.

## ● Evolution by natural selection

Charles Darwin worked out his theory of evolution by natural selection in the late 1830s. He saw that animals and plants all produce too many young and that, in general, only the strongest survive. The features that enable them to survive (bigger teeth, stronger legs, brighter feathers) are passed on to their offspring. In time, the make-up of the whole population may change, or evolve, in the direction of the features that most promote survival at the time. This is natural selection, and Darwin published his famous book *On the origin of species* in 1859.

## ● Mr. Darwin's missing link

What was needed in 1859 to support Charles Darwin's new ideas about evolution was a "missing link" some fossil form that was midway between two quite distinct groups. And in 1861 the most famous missing link of all was found, *Archaeopteryx*. Here was an animal caught halfway between evolving from a reptile into a bird. Perfect timing! In 1861 a complete skeleton of a small dinosaur was found in Late Jurassic rocks in southern Bavaria, but it was a reptile with a difference: it had wings and feathers. Wings and feathers mean it was a bird, but it still had teeth in its jaws, a long bony tail, and claws on its wings.

## ● Pliny Moody and Professor Hitchcock

A schoolboy rejoicing in the splendid name of Pliny Moody came up with an interesting fossil specimen in about 1800. He lived in Massachusetts on his parents' farm, and he spotted a slab of red sandstone that bore rows of small three-toed footprints. Was this Noah's raven, flown out from the Ark? Many more footprints of this kind came to light over the years in the Triassic of New England. Edward Hitchcock, President of Amherst College and State Geologist for Massachusetts, called these footprints "ornithichnites" ("bird footprints") and pictured their maker as a large flightless ostrichlike bird, perhaps ten to thirteen feet tall. He was both right and wrong. The footprints *had* been made by a large biped, but not by a bird.

## ● The first American dinosaur bones

The first dinosaur bones from North America were modest enough: a few teeth collected by an official geological survey team operating in Montana in 1855. They were described in 1856 by Joseph Leidy (1823–91), Professor of Anatomy at the University of Pennsylvania. Small beginnings. However, two years later Leidy was able to report a much more significant find, a nearly complete skeleton of a large plant-eating dinosaur from Haddonfield, New Jersey, which he named *Hadrosaurus*.

## ● A bipedal dinosaur

In interpreting the first complete dinosaur skeleton to be found in North America, Joseph Leidy realized that *Hadrosaurus* was related to *Iguanodon*, but it was younger in age, coming from Late Cretaceous rocks. Most significant was the fact that the skeleton was more complete than anything yet known from Europe. It proved for the first time that this dinosaur at least stood on its hind legs. Owen's Crystal Palace models had already had their day: Leidy's announcement soon showed that many, if not most, dinosaurs had been bipeds. But he didn't get the pose quite right, showing it as rather kangaroo-like (see Chapter 2).

## ● Professor Cope's industriousness

Edward Drinker Cope (1840–97) was a Quaker, taught by Leidy in Philadelphia, and his interests spanned paleontology and herpetology, the study of modern amphibians and reptiles. Cope toured Europe in 1862, to improve his education, and to visit the museums. He taught for a while at Haverford College, Connecticut, but he did not enjoy that and soon became a scientist of independent means, living off his family wealth. In his lifetime, Cope wrote thousands of technical papers on fossil and modern reptiles, and he named over a thousand new species. This is a record that will probably never be beaten (and that is probably not such a bad thing).

## ● Cope's character

Edward Cope was brilliant, aggressive, socially inept, and utterly obsessed. He was impatient and leapt from project to project: this gave him an encyclopedic knowledge of nearly everything in natural history, but he often made mistakes. When Cope died, he donated his body to science, and his skull is sometimes reckoned to be the type specimen of the human species, *Homo sapiens*, his dearest wish.

## The patrician Othniel Marsh

Othniel Charles Marsh (1831–99) was also an enthusiastic paleontologist, but his education followed a more patrician course. He was educated at Yale University, and in the early 1860s also spent three years in Europe. In 1866 his wealthy uncle, George Peabody, was persuaded to donate $150,000 to Yale to enable it to establish a natural history museum. Thus was born the Peabody Museum. Marsh was appointed Professor of Paleontology, an unpaid position, and he relied on his uncle's fortune for financial support for the rest of his life.

## Marsh's character

Othniel Marsh was much calmer than Edward Cope, and he worked slowly and methodically. He was somewhat aloof and overbearing. He employed many "assistants," who in fact did much of his work for him. Whereas Cope wrote every word himself, many of Marsh's famous monographs were ghostwritten by poorly paid aspiring young naturalists he had taken under his wing.

## Cope v. Marsh

The two young paleontologists Cope and Marsh first met during their European tours in the 1860s, and for a time they worked together in a

**Edward Drinker Cope (1840–97), brilliant and mercurial, a prolific and driven scientist.**

**Othniel Charles Marsh (1831–99), more methodical and wealthy, and Cope's sworn enemy.**

friendly manner. Both of them collected and described a range of fossil reptiles and mammals, first from the American east coast and then from the new territories of the midwest. But this friendliness broke down in about 1870, and thus began one of the biggest fights in scientific history, the Great American Bone Wars.

## ● Cope's big mistake

The Cope v. Marsh feud all began because Cope put together a fossil skeleton the wrong way around. Marsh noticed, during a visit to Philadelphia, that Cope had mounted a skeleton of the marine plesiosaur *Elasmosaurus* with its head on the end of the long thin tail instead of at the other end. Supposedly, when Marsh told him of his error, Cope's pride was so hurt that he swore lifelong enmity. The hostility was a constant and major force for the remaining years of their lives. In a way, they seemed to enjoy the fight, the lone mercurial obsessive versus the methodical patrician team leader.

## ● Ambition

Cope and Marsh found enough new dinosaur sites and dinosaur specimens to keep both of them, and armies of assistants, busy for decades. However, the ambitions of both men led them to compete with each other for the best finds. Each had enough money to buy fossils from local collectors and to employ teams of excavators who operated in the midwest. The first excavators were the tough workmen who lived in wild conditions while building the great railroads across America.

## ● Skulduggery

The field men who worked for Marsh and Cope were used to hard work, but they were not trained paleontologists. Nevertheless, Cope and Marsh found that they were resourceful, that they had a good eye for bones, and that they did not have to be paid much. When word came through that a survey team had found some fossil bones, Cope's and Marsh's agents would gather a team of field workers and have them work day and night removing bones at speed. Some of these operations were in dangerous country, and the men were armed against attack. They even spied on each other and muscled in on new prospects.

### Pile 'em high and ship 'em out

The field crews employed by Cope and Marsh in the 1870s and 1880s worked hard and fast, even in the blizzards of winter. It's no wonder that a huge amount of damage was done in these speedy operations. Inferior bones were smashed in the field, and the good ones were hacked out quickly, with none of the careful mapping, strengthening and packaging commonplace today (see Chapter 1). Dinosaur bones were loaded into boxcars and sent east by rail.

### The speed of the scientists

When the latest consignment of huge bones reached them in the 1870s and 1880s, Cope and Marsh fell on the packing cases, tearing them open in a fever to see how many new dinosaurs there might be. They set to describing the new dinosaurs and other wonderful beasts in haste. They rushed their manuscripts to their editors and published new dinosaur names as fast as the presses could roll. Each man had his own pet journal, Marsh the *American Journal of Science*, Cope the *Proceedings of the Natural History Society of Philadelphia*.

### Publish and be damned

At the height of the Great American Bone Wars, Cope and Marsh would often write a brief description of the most remarkable new bones and deliver it to the presses on the same day. It would be typeset, checked, and published in two or three days. Their papers often appear at the ends of the monthly issues of those journals as supplements, tagged on in the day or two before they were posted to subscribers. Even a day could make a difference: a rule of biological nomenclature is that the first name to be given to a new plant, animal, or fossil is the name that everyone uses (see Chapter 1). So if Cope could beat Marsh by a day, his name would stand, the other would fall.

### The positive results of the Bone Wars

Thanks to Cope and Marsh we have many of the most famous North American dinosaurs – *Allosaurus, Apatosaurus* (= *Brontosaurus*), *Camarasaurus, Camptosaurus, Ceratosaurus, Diplodocus, Stegosaurus,* and *Triceratops*. These men also opened up the famous dinosaur sites of Utah, Colorado, the Dakotas, and Montana. At the same time their collectors investigated younger rock layers and turned up spectacular finds of fossil mammals and birds.

### ● The mass burial at Bernissart

Many new dinosaurs were found in Europe in the late nineteenth century, but one attracted a great deal of attention. This was the discovery of the mass grave of dozens of skeletons of *Iguanodon* found at Bernissart in Belgium. In 1877 coal miners working a deep shaft, more than 1000 feet below the surface, came upon large bones in the roof of a cutting. The mining company contacted scientists in Brussels, who were able to identify the remains as belonging to *Iguanodon* by examining the isolated teeth. Normal mining operations were stopped, and paleontologists from the Royal Museum of Natural History in Brussels moved in and supervised the careful excavation of 39 skeletons of *Iguanodon*, most of them essentially complete.

### ● Louis Dollo's life work

When Louis Dollo (1857–1931) was appointed assistant at the Royal Museum of Natural History in Brussels in 1882 he was given the job of sorting, preparing, and describing the astonishing Bernissart dinosaur collection. The job took him most of the rest of his life. He supervised the cleaning of the skeletons, which was not an easy task since the bones were damp and cracked. The museum technicians had to use a terrifying cocktail of varnishes and glues to strengthen the bones, and this has led to endless conservation problems ever since. For the first time in Europe, Dollo was able to reconstruct some complete dinosaur skeletons in their natural pose, efforts that rivaled those of Leidy, Cope, and Marsh in North America.

### ● The troublesome nose horn

Louis Dollo was able to solve a long-standing problem about *Iguanodon* in the 1880s. A strange conical-shaped bone had been found in the 1830s but no one knew where it went. Mantell thought it was a nose horn, and indeed Owen reconstructed *Iguanodon* in the 1850s in his Crystal Palace models with the bone mounted on the snout (see page 29). The new skeletons from Bernissart showed that the mystery bone was actually a specialized thumb claw, used presumably for defence or in fighting for mates.

### ● Andrew Carnegie's millions

Andrew Carnegie, a Scotsman who had made his fortune in America in the steel industry, was persuaded to fund a museum in Pittsburgh, known, oddly enough, as the Carnegie Museum. He was fascinated by amazing new finds from quarries in the Late Jurassic Morrison Formation in Colorado. One of the best finds from the quarry was a complete skeleton of *Diplodocus*, which was named *Diplodocus carnegiei* (H'mm. Money can buy you anything.) This greatly pleased Carnegie, who had always said he wanted his museum crews to find a dinosaur 'as big as a barn'. He had molds made of every bone and gave away free plaster casts of this huge dinosaur to museums in London, Paris, Frankfurt, Vienna, La Plata, and Mexico City.

## ● Off to Africa

While dinosaurs were found mainly in Europe and North America in the nineteenth century, attention shifted to the southern continents in the twentieth. A promising new site was discovered by a mining engineer who was working in Tanzania, then German East Africa. He brought back tales of huge bones lying around in the scrubland near Lindi, at a site called Tendaguru. Paleontologists visited the site and saw its huge potential. They went back to Berlin and launched an appeal for funds, raising the huge sum of 200,000 marks from the government, from various scientific organizations and from wealthy donors.

## ● The biggest dinosaur expedition

The biggest dinosaur expedition ever ran from 1907 to 1912 in German East Africa. In charge was Dr. Werner Janensch of Berlin. In the first season at Tendaguru, 170 native laborers were employed, and this number rose to 400 in the second season and 500 in the third and fourth. These huge numbers of workers were accompanied by their families, so the German dinosaur expedition at Tendaguru involved an encampment of 700 to 900 people. Hundreds of huge bones were carried to Lindi, shipped to Berlin, cleaned up, and mounted. The Berlin specimen of *Brachiosaurus* is the largest complete dinosaur skeleton in the world.

## ● The dinosaurs of Mongolia

Asiatic dinosaurs really hit the headlines in 1922, 1923, and 1925. An American expedition led by Roy Chapman Andrews, a paleontological technician, returned from Mongolia with spectacular dinosaur specimens: skeletons of the small ceratopsian *Protoceratops* with nests containing eggs, and the extraordinary slender meat-eaters, *Saurornithoides*, *Velociraptor*, and *Oviraptor*. Andrews and his team had

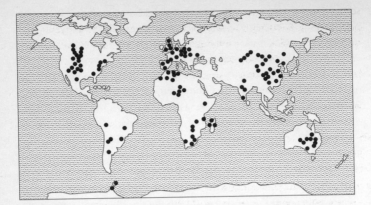

**Map of the world today, showing major sites of dinosaur finds. Most discoveries have come from classic sites in Europe and North America, largely from those opened up in the nineteenth century. Africa, South America, Asia (especially India, China, and Mongolia), and Australia were opened up in the twentieth century. In 1988 dinosaur fossils came to light in Antarctica.**

been sent out by the American Museum of Natural History to look for the ancestors of humans. They failed to find them, but their dinosaur collections were stunning. Work in Mongolia continues to this day, and dozens of spectacular Late Cretaceous dinosaurs have come to light, some of them unique to the area.

## ● Dinosaur recession

It may seem amazing today, but dinosaur paleontology went through a recession during the middle decades of the twentieth century. Major expeditions were run by Russian, Chinese, and Polish teams, but very little was going on in Europe or North America after those great Mongolian expeditions in the 1920s. One of the few high points was the discovery of skeletons of *Deinonychus* in the Mid Cretaceous of Montana by John Ostrom in 1964. And these Montana finds perhaps launched the latest dinosaur craze.

## ● Dinosaur renaissance

Bob Bakker in 1975 penned an article in *Scientific American* entitled "Dinosaur Renaissance." He was drawing attention to the fact that something amazing was happening in the field of dinosaurian paleobiology. Suddenly dinosaurs were sexy again, and bright young geologists and biologists wanted to work on dinosaurs. Publications about dinosaurs were burgeoning. Museums were hiring paleontologists once again; expeditions were setting out and finding new species of dinosaurs all over the world, including Antarctica. What had happened?

## ● Bob Bakker

Bob Bakker was more often wrong than right, but he had a huge influence in the 1970s. He was full of exciting ideas about dinosaurs: they were warm-blooded (see Chapter 6), and they ran fast (see Chapter 5), they had complex social behavior (see Chapter 7). He was also a superb artist, who drew haunting, dynamic images. His 1969 sketch of *Deinonychus* was a classic (see page 35). His publications about dinosaurian physiology and behavior stimulated and infuriated palaeontologists in equal measure. Debates raged around all aspects of dinosaurian paleobiology, and the debates were carried out in public as well.

## ● The 1990s

The renewed interest in dinosaurs grew in the 1980s and 1990s. Two further controversies pumped it up further. In 1980 Luis Alvarez, a two-time Nobel Prize-winning physicist, and his colleagues proposed that the dinosaurs met their end by the impact of a huge asteroid on the earth. This reopened the debate about the extinction of dinosaurs, a debate that still rumbles on (see Chapter 10). In the 1980s and 1990s paleontologists were also experimenting with new methods for drawing up evolutionary trees. One of the big issues was the origin of birds: birds came out time and time again as living dinosaurs (see Chapter 9). Dinosaur palaeontology has never been so healthy as it is now: there are more dinosaur paleontologists in jobs around the world than ever before. They are finding ever more species (twenty new dinosaurs were named in 1999 - see the Appendix), and more and more people of all kinds are fascinated to find out more about what we know.

CHAPTER FOUR

# DINOSAUR DIETS

Dinosaurs ate to live. Mostly, we have to guess what they would have eaten, but the skeleton is full of clues. For example, tooth shape is a good guide to diet: plant-eaters (herbivores) have blunt hard-wearing teeth, while flesh-eaters (carnivores) have pointed teeth. Herbivores are fat, carnivores thin (think about why that might be). But in some cases there is direct evidence of dinosaurian diets: gut contents, fossil feces, even bones with tooth marks. So it's not all guesswork by a long shot.

It's important to follow through in detail just how the paleobiologists make their decisions. In one sense it's all "speculation," but then so is pretty well all of science when you think about it. Astronomers study the nature of planets and stars they've never set foot on: is this wild and reprehensible behavior? Molecular biologists can never put their finger on the chemical structures they study, but should they be stopped from their wild speculations? Reconstructing dinosaurs is no different.

## ● Habitats and communities

It's useful to think a bit about some basic ecology before we go on. When biologists look at a modern situation, they think in terms of a habitat (the physical setting where the plants and animals live) and a community (the assemblage of plants and animals that live together). Communities have evolved, often over millions of years, so that each species has its place, and all species depend on each other. Disturb the community in some way, and everything may be thrown off balance. This is a big political discussion point now, of course, as we look at the effects of human activities on the natural world.

## ● Energy transfer

Starting at the beginning, as one should, the whole of life can be thought of as energy processing. Energy comes from the sun. It is captured by green plants that photosynthesize and create leaves and stems and tree trunks. The energy is fixed in the form of carbon and other atoms. Carbon, as plant material, is eaten by herbivores, which are then eaten by carnivores, and then everything goes back into the soil through death and decay. Energy and carbon then cycle on endlessly through the physical and the living worlds.

## ● Food chains and food webs

Energy transfer moves through successions of consumers. So it's useful to think of the relations among all the plants and animals in a community in terms of their feeding relationships. In other words, who eats what? Plants capture the energy, and then plant-eaters small and large devour the leaves. The herbivorous animals, ranging from insects to deer and elephants, are then eaten by flesh-eaters. Dragonflies are key carnivores that eat herbivorous insects. Lions and tigers eat the bigger herbivores. Then, eventually, all animals die, and their bodies are recycled by decomposing microbes back into the soil. So we can think of a food chain running like this: grass → rabbits → foxes → decay in soil → grass and so on. This is very simple, and of course the real situation is usually more like a complex web with arrows crisscrossing.

## ● Fossilization

We don't know every dinosaur that ever lived, but we do know some dinosaurian communities in quite a bit of detail. Here and there, conditions were especially favorable to the preservation of fossils. Maybe there were some lakes and rivers around which a wide range of plants and animals lived, and perhaps the plants, insects, and fishes sank to the bottom of the lakes and were buried in fine black mud. Nothing would disturb their remains, and we may find exquisite fossils, with every detail of an insect's wings and the cells in a leaf. Bigger beasts would be buried in the lakes, rivers, or soils, and their skeletons may, sometimes, be quite complete. So in the end there are a few dozen dinosaurian communities, from different parts of the world, that we know in intimate detail, from the smallest microbe and insect to the biggest looming sauropod dinosaur.

## ● Ecology of the Morrison Formation

The Morrison Formation in the midwestern United States is one example of a well-studied rock unit. This unit of rocks – that's what a "formation" is – extends for hundreds of miles over parts of seven states: New Mexico, Oklahoma, Colorado, Utah, Montana, South Dakota, and Wyoming. The Morrison Formation is dated as Late Jurassic, perhaps some 150 million years old. What's special about it, apart from the hundreds of astonishing skeletons excavated there since the days of Cope and Marsh (see Chapter 3), is that it has yielded fossils of all the other plants and animals that lived at the time.

## ● Plant food

The plants of the Morrison Formation were quite different from what is seen around today. For starters, there were no flowering plants, and that means no grass, no deciduous trees, no bushes or shrubs, vegetables or fruits of the kinds that feed us and our livestock. Indeed, this was true for all of the Triassic and Jurassic, and for the first part of the Cretaceous – indeed, pretty well the first 100 million years of the existence of dinosaurs on the earth. So what plants were there?

## ● Ferns and horsetails

Dinosaurs would have seen (and eaten) some plants that would look familiar to us. There were ferns and horsetails, for example. But at that time these rather primitive plants probably occurred much more widely. Remember, there was no grass, so ferns and horsetails would have

The cycadeoid *Cycadeoidea*, superficially like an overgrown pineapple, was key dinosaur fodder.

spread out over damp ground anywhere they could. As *Diplodocus* stomped around, it was ferns and horsetails it squashed beneath its great feet. Ferns and horsetails had been around on the earth for a long time – in fact, they had been some of the first plants to become established on land 400 million years ago in the Devonian period, when life moved decisively on to land. But the dinosaurs didn't know this. They simply chomped their way through acres of the stuff. And the ferns and horsetails always grew back, because these plants are linked by underground rooting bodies – rhizomes – as any gardener can tell you (much to his annoyance).

## ● Tree ferns and club mosses

Some of the dinosaurs' favorite fodder would look less familiar to a modern botanist than the ferns and horsetails. Tree ferns were relatives of ferns, but they grew a trunk and reached modest heights, as the name suggests. The club mosses were not really mosses. Called more properly lycopods, these plants grew to the size of small bushes. They had a central stem and a crown of succulent leaves. These groups also had to grow somewhere near water. What of the drier habitats?

## ● Conifers

Dinosaurs lived mainly in rather hot and dry habitats. Away from rivers and lakes, there were dry upland plants. Most important were the conifers, and many of these would look familiar. Indeed, you've seen the great forests of monkey puzzles and proto-firs and spruces shown in *Walking with Dinosaurs*. Conifers were the dominant trees during the Mesozoic. They can withstand dry conditions, since their leaves are hard, waxy structures. Think of pine needles: they are hard and thorny to the touch. They contain much less water in their cells than the leaves of, say, an oak or a beech. Tough leaves required tough herbivores, of course. Many dinosaurs had a hard time cracking their way through mouthfuls of pine needles, but that was all they knew, so they probably had no complaints.

## ● Maidenhair trees, cycads, and cycadeoids

One or two plant groups filled up the spaces among the conifers and the ferns. The maidenhair tree, or ginkgo (note the spelling – people often misspell it "gingko"), is nowadays restricted naturally to China. It's a beautiful tree, with a smooth, delicate brownish-purple trunk and festoons of delicate green leaves. It has now been spread worldwide and is a favorite tree in many towns and gardens. In the Mesozoic, ginkgos were much more diverse. Cycads and cycadophytes were related plants, which grew much closer to the ground. They either had a straight trunk or a ball-like trunk structure out of which sprouted a rosette of fern-ike leaves (see page 60). Cycadeoids are extinct now, but cycads are still found in the tropics today.

## Flowering plants

Most dinosaurs never saw a true flower or a deciduous tree. The first flowering plants – or angiosperms, as they are known botanically – appear in rocks of Early Cretaceous age, perhaps 120 million years old. It's still unknown when the angiosperms first appeared, and paleobotanists are constantly tussling over ever older reports of supposed Triassic and Jurassic angiosperms. But none of these early records has been universally accepted.

## Success of the angiosperms

The first flowering plants were relatives of the modern magnolia, roses, vines, and other forms of that kind. Angiosperms have flowers, critical in their breeding cycles. They were good news for the Cretaceous dinosaurs, since they could be grazed down to the ground and they would spring up again from underground rhizomes or from seeds. The key to their success was their seeds, which could be produced in vast quantities and were spread rapidly over large areas, often by the wind.

## Bugs, beetles, and flies

The Morrison Formation landscape was abuzz with flying creatures, bugs, beetles, and flies abounding. Dung beetles had never had it so good: the droppings from one dinosaur would keep a thousand dung beetles happy for weeks. Cockroaches bustled around in the leaf litter, helping the decomposition cycle along. Carrion beetles and flies fed on decaying dinosaur carcasses. Leaf beetles chomped the leaves. Mosquitoes knifed through the skin of the reptiles and sucked their blood. Dragonflies soared and whirled at sunset doing the decent thing, putting thousands of mosquitoes out of their misery every evening. But there weren't any ants, bees, wasps, or butterflies. These groups appeared only later, in the Cretaceous, since they evolved hand in hand with the angiosperms. This is a famous example of co-evolution: the flowering plants needed polinating insects, and the polinating insects needed the flowering plants.

## Fishes in the lakes

The Morrison lakes and rivers were stuffed with fish. But there were no salmon or trout. They were certainly fishy fish, not the primitive heavily armored fishes of the Devonian, 350 or 400 million years ago. But they were still primitive to our eyes. They had heavy bony scales, unlike the thin transparent scales of most modern fish. The Morrison Formation was timed on a cusp in evolution. A major new fish group, the teleosts, had appeared earlier in the Jurassic, but they were not yet hugely diverse. Teleosts had thin scales, they were fast-moving and they had highly adaptable jaws that could suck, nip, snatch, and grab food quickly and effectively. Today, teleosts dominate lakes and oceans: they include everything from goldfish to gurnard, from tuna to turbot, from salmon to sea horse. There were sharks in the Mesozoic, of course, and even some freshwater sharks that prowled the Morrison waterways.

## ● Other small creatures

The shores of the Morrison rivers and lakes were prowled by some pretty familiar little beasts. There were three or four species of lizards, not so very different from modern lizards. But no snakes. Snakes appear only in the Early Cretaceous. There were turtles, some of them smallish terrapin-like creatures that wallowed on the bottoms of the warm, murky ponds, others high-shelled land-livers that stuffed themselves with cockroaches and crickets. There were crocodiles, mostly quite modest in size, that probably fed mostly on the scaly fishes. So far, no birds have yet been found in the Morrison Formation, although some isolated scraps of bone might come from birds. Some day, convincing birds may be found: after all, the Morrison Formation is not much older than the rocks in Bavaria that produced *Archaeopteryx* (see Chapters 3 and 9).

## ● Mammals

Oddly modern-looking among the dinosaurs of the Morrison Formation were the mammals. A few nice jaws of small shrewlike mammals have been found in Morrison sediments. These were tiny beasts, only a few inches long, and probably the dinosaurs hardly knew they existed: there were three reasons for this. First, the mammals were just so tiny: *Diplodocus* could have stomped on a hundred and not realized it. Secondly, the Morrison mammals, like most modern mammals were nocturnal. They knew that it was safest to hunt at night, and their warm-bloodedness made this possible. While the crocodiles and dinosaurs were snoozing, the mammals flitted around snatching up earthworms and small bugs. Thirdly, the dinosaurs probably couldn't see well enough to pick up such insignificant little specks. But, of course, these insignificant little beasts were our ancestors, so we mustn't really disparage them.

## ● Pterosaurs

Soaring over the Morrison Formation trees were some familiar forms. At least, they were familiar because we have seen them in so many paintings and Hollywood films. The Morrison lands were overflown by pterosaurs. It is interesting that very few pterosaur bones have actually been found yet in the Morrison Formation rocks, but we know that pterosaurs were diverse and abundant from findings in other Upper Jurassic rocks, especially in Germany. Beside an *Apatosaurus* or *Brachiosaurus*, the Morrison pterosaurs were tiny, no more than two feet in wingspan. The pterosaurs fed on insects or fish.

## ● Small plant-eating dinosaurs

The Morrison dinosaurs are what many people want to know about (see page 67). There were some thirty species, a good diverse assemblage, and this is probably something close to the total diversity that actually existed. Small plant-eaters included the bipedal ornithopods *Othnielia*, *Drinker*, and *Dryosaurus* and slightly larger was *Camptosaurus*. These all had rows of ridged teeth along their jaw margins, so they were adapted to a tough diet, perhaps feeding on the low-growing ferns, horsetails, club mosses, and cycads. They were also all fast-movers, with long slender legs and stiff balancing tails. But at their size, from three to sixteen feet long, they couldn't wait around if a slobbering *Allosaurus* came on the scene.

## ● The plated stegosaurs

The most fascinating Morrison plant-eater was *Stegosaurus*, of which there were perhaps three species. *Stegosaurus* is probably one of the top five favorite dinosaurs of all time. Everyone can picture its symmetrical profile – high humped back, low short neck and tail, and the row of great bony plates set along its back. *Stegosaurus* is famous for many things – its plates, its minute brain, its spiked tail. The plates may have been used in temperature control (see Chapter 6) or for scaring enemies (see Chapter 8). The spiky tail was for whacking predators. The tiny brain was supposedly the smallest (relative to body size) of any dinosaur. But in the tooth department, *Stegosaurus* was well-endowed. It had rows of broad, slightly pointed shearing teeth along the jaw margins. It could only feed close to the ground, and probably joined the small ornithopods in chomping ferns and cycads.

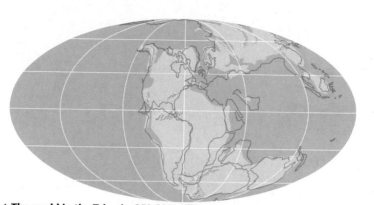

▲ The world in the Triassic, 250-205 million years ago, one super continent.

▲ By Jurassic times, 205-145 million years ago, the North Atlantic was opening.

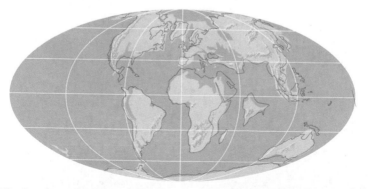

▲ The Cretaceous world (145-65 million years ago) began to look more familiar.

◀ **The duckbilled dinosaur or hadrosaur *Edmontosaurus* takes a drink. In the Late Cretaceous, hadrosaurs were hugely abundant and fed on tough vegetation.**

▲ **One of the first plant-eaters, the prosauropod *Plateosaurus* had large hands, with massive thumb claws, well adapted for raking and grasping vegetation.**

▶ **The biggest plant-eater, *Brachiosaurus*, from the Late Jurassic of North America. With their long necks, these monster sauropods could feed in higher trees than any other animal.**

▲ The giant pliosaur *Liopleurodon*, from the late Jurassic of Europe, fed on other sea reptiles such as the ichthyosaur *Ophthalmosaurus*.

◄ The pterosaur *Tapejara* was a huge flier with a wing span of about sixteen feet. It fed on fish scooped from the sea.

▶ One of the best-known dinosaurs, *Iguanodon*, from the Early Cretaceous of Europe, is known from dozens of skeletons. *Iguanodon* was a ponderous animal that could move equally well on all fours or on its hind legs only.

## ● The behemoths

Higher plants were consumed by the sauropods, the giants of all time. At 55 tons or more, the Morrison Formation sauropods were an impressive bunch: *Apatosaurus, Barosaurus, Brachiosaurus, Camarasaurus, Diplodocus, Haplocanthosaurus, Seismosaurus, Supersaurus,* and (you've guessed it!) *Ultrasaurus.* They varied in relative proportions, head shape and tooth type, but they were all huge. There is a running debate about how much a sauropod ate every day – whether they were warm-blooded or cold-blooded (see Chapter 6), though, it was a huge amount. And the big question is: how on earth could they vacuum up enough plant food with that tiny mouth and slurp it down their necks? No one knows, but they clearly did, and they were obviously good at whatever they did do.

## ● Small flesh-eaters

The smallest Morrison dinosaurs were carnivores. *Ornitholestes,* for example, was six feet long. A small birdlike biped, *Ornitholestes* scuttled rapidly from bush to bush snatching up lizards and mammals when it could, or sometimes, a juicy dragonfly or cockroach. A close relative was *Coelurus*; indeed, paleontologists still debate whether they are distinct forms or whether they might be one and the same.

## ● Big flesh-eaters

The big flesh-eaters of the Morrison Formation are quite familiar animals. Dozens of skeletons of *Allosaurus* have been excavated. Here was a 400-foot-long beast, a classic carnivorous dinosaur: long powerful legs, a heavy tail for counterbalancing, small arms but with powerful hands, and a deep skull with powerful jaws. Another large flesh-eater was *Ceratosaurus,* superficially similar to *Allosaurus* but with a pair of pointed bumps ("horns") in front of its eyes. There were a few other less well-known theropods in the Morrison. (Theropods include all meat-eating dinosaurs.) *Allosaurus* and *Ceratosaurus* fed on the plant-eaters like *Dryosaurus* and *Camptosaurus,* and they would certainly have tackled *Stegosaurus*. But what about the huge sauropods? Did they have any predators?

## ● Immune from attack?

Paleontologists have often wondered why the sauropods didn't seem to have any way of defending themselves. They were the original

gentle giants, 55 tons of floppy flesh and not a spine, horn, or a serious
armor plate. But maybe, like modern elephants, they were immune
from attack. It would have taken a dozen *Allosaurus* to gang up against
one *Apatosaurus* adult and bring it down. They probably wouldn't have
been able to pull it off. But the baby sauropods were certainly at risk.
Huge dinosaurs laid big eggs, but not in proportion to their body size.
(If a 12-inch-long hen lays a three-inch-long egg, why didn't a
100-foot-long *Diplodocus* lay a 23-foot-long egg, as big as a Rolls-
Royce? Think about it. The answer is in Chapter 7.) Baby *Diplodocus*
were probably only six feet long when they hatched, so they were
prime targets of all the flesh-eating theropods in the Morrison.

## ● The Morrison food web takes shape

It is possible to draw up a pretty detailed food web for a fossil community.
Now we have all the plants, fish, bugs, small creatures, and the thirty or
so dinosaurs (see page 67) sorted out, we can do this for the Morrison
Formation. Plants go at the bottom, and then the web builds up through
a number of interlinked chains, through fish, insects, and plant-eating
reptiles. Some chains are short (e.g. trees → adult sauropods), but
others can be quite long (e.g. water plants → fish → crocodiles →
*Ornitholestes* → *Allosaurus*). The more you think about it, the more
tangled the whole food web becomes. But then that's nature. Life,
evolution, co-evolution, and natural communities are highly complex.

## ● What is the evidence for diet?

We've already seen that paleontologists can disentangle the probable
diets of individual dinosaur species. Clearly, dinosaurs could eat only
what was around in the neighborhood. And we've already heard that
the teeth, jaws, and skeleton in general can give clues. But how?

## ● Teeth of meat-eaters and plant-eaters

The teeth of carnivorous dinosaurs are instantly recognizable (see page
68). They are curved and pointed in side view, and rather flattened.
Plant-eating dinosaurs didn't ever have these kinds of teeth. Theirs may
be rather feeble little pencil-like structure, or sort of leaf-shaped in
side view, with a narrow ridge fore and aft. In some forms that
specialized on especially tough vegetation, the tooth has extra ridges
and grooves to give it strength.

**The dinosaurs of the Morrison Formation, c.150 million years ago, of midwestern North America. Plant eaters and meat eaters are distinguished, and all are drawn to scale. *Allosaurus* could eat *Stegosaurus*, but what could eat the giant sauropods?**

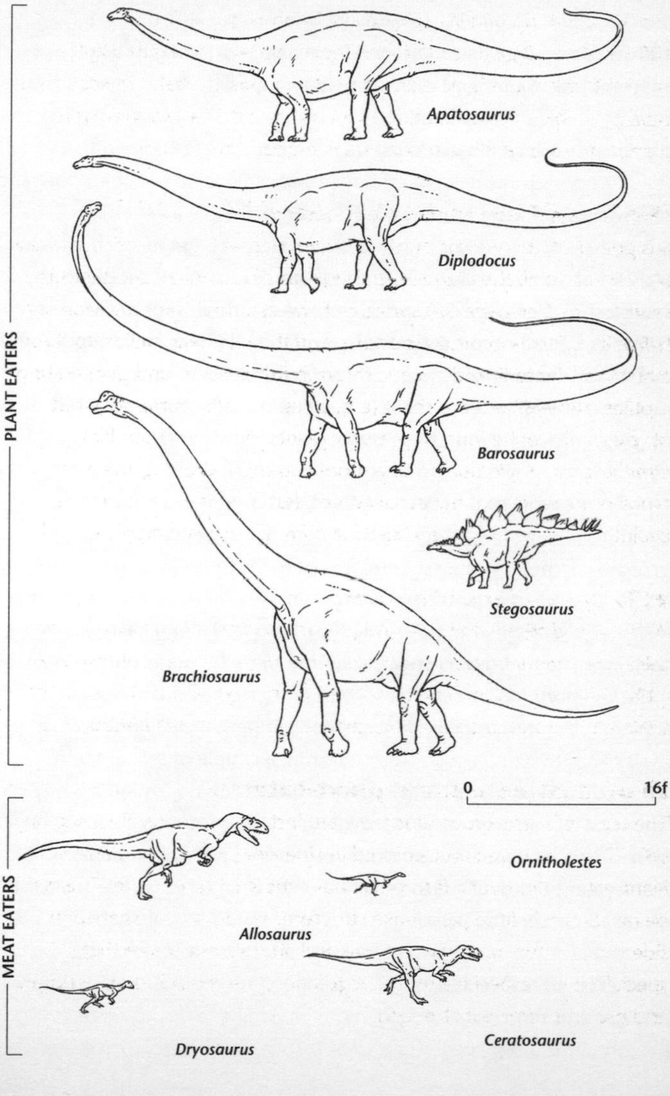

PLANT EATERS

*Apatosaurus*

*Diplodocus*

*Barosaurus*

*Stegosaurus*

*Brachiosaurus*

0      16f

MEAT EATERS

*Ornitholestes*

*Allosaurus*

*Dryosaurus*

*Ceratosaurus*

Plant eater – *Diplodocus*　　　　Meat eater – *Allosaurus*

**A plant eater and a meat eater, showing their different tooth shapes, different jaw shapes and different skull shapes.**

## ● Pencil teeth of the sauropods

The sauropods must have been huge food processing factories, downing several tons of food a day. And the food didn't have a very high nutritional value – you have to eat a lot of ferns to get enough protein and carbohydrate. Amazingly, the sauropods had pretty feeble teeth, and most of them had only a bunch of teeth at the front of the jaws. *Diplodocus*, for example (above), had only about forty teeth altogether, each about the size and shape of a pencil, and pointing slightly forwards, giving it a bit of a goofy expression. *Diplodocus* could certainly nip off some leaves with these teeth, but it couldn't chew.

## ● The hadrosaurs and their 2000 teeth

The most amazing dinosaurian dentitions are to be seen in the hadrosaurs, the duck-billed dinosaurs of the Late Cretaceous. These would have been a dentist's dream: we have a mere 32 teeth in all; the hadrosaurs had a full set of up to 2000, with 500 along each side of each jaw. The hadrosaurs specialized in especially tough vegetation: some skeletons have been found with pine needles in the stomach area. But it's more to do with their reptilian heritage. Dinosaurs, like sharks and modern reptiles, did not grow a couple of sets of teeth during their lives. Their teeth kept growing continuously. So if a few teeth got worn down or were broken off, a new set would soon grow up to take their place. Inside the jaws of the hadrosaurs were banks of replacement teeth, all ready to move up to the jaw edge, grow into place, and come into action.

## ● Jaws of the plant-eaters

Plant-eating dinosaurs had powerful jaws. The jaw lines were straight, and often carried a row of forty to fifty teeth. The secret in feeding is the location and strength of the bite point. Our strongest bite point is at the front of the mouth, so we can bite off a tough bit of toffee or tear meat from a barbecued rib using our incisor teeth. In herbivorous dinosaurs, the maximum bite point could be at the front of the jaws, but it was often a broader area in about the middle of the tooth row. To achieve this, the hinge point – the point where the jaw attaches to the skull – was set lower than the tooth row (see page 68). This gave extra power to the jaws and a broad distribution of the bite point.

## ● Jaw muscles

Move your jaws about and place the palms of your hands on each side of your head, just over the ears: as you chew, you can feel the muscles moving over the side of your skull. Things were quite different in dinosaurs, which kept their jaw muscles tucked away neatly inside the skull. Of course, mammals are odd, because our braincase has swollen so large that the jaw muscles have to sit on the outside. The main jaw muscles are for snapping the jaws shut, then for making a back-and-forwards motion, and then for sideways movements. There is a much weaker muscle behind the jaw joint for opening: it can be small, – gravity does most of the work. Just watch a child concentrating: his mouth dangles open simply under the influence of gravity.

## ● Gathering in the food

Did the herbivorous dinosaurs simply use their mouths to gather in the food? In the quadrupedal forms, yes. But the bipeds had their arms to help. Some, such as the prosauropods like *Plateosaurus* from the Late Triassic, had strong arms and broad hands. The prosauropods had a great flattened, hooked thumb. Perhaps this was used for raking branches together to speed up the feeding process.

## ● Reaching high in the trees

Many dinosaurs could reach high into the trees. The bipeds could certainly stretch up on their tiptoes. Could *Stegosaurus* tip back on its haunches and reach higher? This is debated. It's even possible that the huge sauropods could extend their size advantage enormously by rearing up. On all fours, *Diplodocus* might have been able to reach up to 16 to 20 feet or so. *Brachiosaurus*, with its much more vertical neck, might have managed 33 to 50 feet.

## ● Could the sauropods rear up?

It has often been debated whether the giant sauropods could have tipped back on to their bottoms, perching on a tripodlike arrangement of their hind legs and tail. If it was possible, they could have achieved a height of nearly 65 feet. That would have been truly awesome: a rearing *Diplodocus* could have peeked in at the sixth floor of a modern tower block. This rearing-up behavior is shown in *Walking with Dinosaurs*, but it's still very much debated. There is some evidence in the tail bones. About a third of the way along the tail, some sauropods have unusually shaped bones beneath the tail, perhaps part of a propping point to create the tripod effect. But then it's scary to imagine a 55-ton animal (that's fifteen times the weight of an elephant) hoisting itself up just to grab a bunch of tasty leaves that would otherwise be out of reach.

## ● Dinosaur tongues

Dinosaurs had tongues, and they clearly used them for feeding (as well as for bellowing and roaring, no doubt). But how can anyone know anything about the tongues of extinct animals? Of course, there are no fossil tongues. The tongue is such a succulent piece of flesh that it would be snatched out of any carcass within moments of death. But the tongue is fixed into the back of your throat by special small bones called the hyoid bones. (These are the bones that forensic surgeons look at if someone has been killed in mysterious circumstances; they are broken in strangling.) Hyoid bones are narrow rods, and they have been found in many extinct reptiles. So dinosaurs obviously had tongues, and powerful tongues at that. Perhaps they could have poked their tongues out and wrapped them around bunches of leaves or ferns, just as a cow or a giraffe can do.

## ● Dinosaur cheeks

If you have ever watched a reptile or a bird feeding, it wastes a huge amount of food. It grabs a piece of vegetation or meat, snaps the jaws shut, and swallows. Anything to either side just falls on the floor. Mammals, supposedly superior in every regard, at least are tidier eaters. They have loose lips and cheeks. The bits of food out on either side are caught and chewed and swallowed. This takes some learning of course: you would hardly know that your kids had cheeks for this purpose! Some plant-eating dinosaurs had cheeks too. The ornithopods, ceratopsians, stegosaurs, and ankylosaurs had their teeth set inwards on the jaw margin, leaving a bony ridge. Almost certainly they had fleshy cheeks. That's why their mouths look small and neat, while crocodiles and flesh-eating dinosaurs have huge open grins.

## ● Ornithopod chewing

It's a well-known fact that only mammals can chew their food. As a child, you were encouraged to chew properly or you would get stomach ache. Well, birds and reptiles just gulp their food down without any chewing at all. This is because their jaw joints are simple hinges. The jaws can only go up and down. Some extinct reptiles could move their jaws back and forwards a bit, so they could saw their way through a piece of tough food. But the ornithopods could chew. Dinosaurs like *Camptosaurus, Iguanodon,* and the hadrosaurs had a bizarre chewing mechanism, quite unlike ours. In mammals, chewing is achieved by having a somewhat ball-like jaw joint, so the lower jaw can rotate from side to side and back and forwards. Try it. In ornithopods, the upper jaw was a bit like a flap, with an extra hinge running from back to front of the long snout. When it closed its jaws, the upper jaw flaps were pushed out a bit. When it opened its jaws, the flaps moved back inwards. The net result was an in-and-out sideways movement as the jaws went up and down, a kind of chewing.

## ● Stomach stones

Do birds and modern reptiles suffer from terrible stomach ache and gas? After all, they don't chew their food. On the whole, they apparently do not. Likewise, too, the dinosaurs. For many birds the solution is to swallow grit, which is stored in the gizzard, a baglike structure halfway between the mouth and the stomach. Hens are always pecking about for grains of sand. When they swallow the seeds and grain, which are their food, whole, these are ground up in the gizzard before going into the stomach as a mash. We don't know for sure, but it seems almost certain that many herbivorous dinosaurs also used this system. But for them, grains of sand would be no use. They chose pebbles one to four inches across and stored them in their stomachs, or in a gizzardlike structure. The gulped-down bits of leaf and twig would be mashed up. The evidence is that many dinosaur skeletons have been found with masses of unusual polished pebbles inside the ribcage. These are called gastroliths ("stomach stones").

## ● Barrel guts

Cows are round in cross section; cats and dogs are narrow. Just look at a cow from behind (stand back), and you'll see. This is because herbivores have very long guts, while carnivores do not. Humans are omnivores ("all-eaters"), so we fall in between. Meat is easy to digest, since it breaks down into proteins, fats, and carbohydrates soon after swallowing. But plants are not. The fibers are made from lignin, an indigestible material. Indeed, most herbivores today have intestinal floras of bacteria to help break down and ferment the plant material. Herbivore = long winding gut = barrel-shaped body + lots of gas.

## ● Coprolites: Mesozoic manure

One of the most intriguing occupations is to be a coprologist. This is someone who studies coprolites ("poop stones"), so you could probably think of some vernacular alternatives to the rather grand term "coprologist." We've already met the redoubtable Dean William Buckland, who in 1824 named the first dinosaur *Megalosaurus* (see Chapter 3). He founded the science of coprology, and he used to enjoy shocking his more delicate acquaintances with his calculations of the volume of dung produced each year by the extinct Mesozoic monsters. Here and there, paleontologists have disinterred dinosaur droppings. Some are long and sausage-shaped, others large and round (compare dog and cow). The coprologist, of course, delves into his fossils, dissecting coprolites and exploring, under the microscope, what the animal had just eaten. Splendid.

## ● Cow-pats as big as squash courts?

How big was a *Diplodocus* dropping? If a cow produces something up to a yard across on a good day, what about a sauropod? Hard to say, but there are some clues. For years, paleontologists have noticed that dinosaur bones are often found in so-called "plant debris beds," layers of rock, maybe only an inch or two thick, that contain masses of cut-up pieces of leaf and twig. Are these sauropod droppings? I am convinced that they are. Unlike the better-formed dropping of a carnivorous dinosaur, the herbivore pat is a liquid pool, probably 16 to 33 feet across. Much of it would disappear as sediment covers it over, so all that is left is a mass of undigested plant debris. So stand clear when *Diplodocus* lifts its tail. It has a ton or two to drop, and probably enough gas to fill a hot air balloon.

## ● Fangs of the flesh-eaters

The flesh-eating dinosaurs all had curved, flattened, sharp-tipped teeth (see page 68). At the front and back is a sharp ridge running from the root of the tooth to its tip. In cross section, if you slice a theropod tooth straight through, it is shaped like a human eye as drawn by the ancient Egyptians, sort of oval but with neat symmetrical points at each corner. The fore and aft ridges show up as the points in the cross section.

## ● Serrations like steak knives

Close-up, the stabbing teeth of theropod dinosaurs were fiendish. Most of them had finely serrated front and back edges, maybe two or three zigzags per millimetre. As the tooth came into use, these serrations wore down gradually. Eventually they would be blunted, but by then the tooth had had its day and it would be shed, and a new fiendishly sharp specimen would move into place. An ever-ready, self-sharpening set of steak knives.

## ● You can struggle, but you can't escape

Why did theropod teeth curve backwards? Why would they not be straight up and down, for example? This is true of other flesh-eaters. Look at the teeth of a shark: they point back down the throat. It's even more obvious in snakes, which have wicked thin tubular teeth that point right back. When any of these predators snatches a prey animal, the victim will clearly struggle. With backwards-pointing teeth, all the struggling does is to move the hapless little herbivore further and further down the throat. A simple and satisfactory little scheme.

## ● Toothless carnivores

Some of the most advanced flesh-eating dinosaurs had no teeth at all; the ornithomimosaurs, for example, so-called "ostrich dinosaurs" of the Late Cretaceous of North America and central Asia, had sharp horn-covered jaw margins. (Admittedly, the oldest ornithomimosaur from the Early Cretaceous of Spain still had tiny teeth, which were lost in later forms.) Other groups, such as the oviraptorosaurs, were also toothless; others, such as the troodontids, had tiny teeth. This seems odd, in such apparently "advanced" theropods. But remember that vultures and eagles haven't got a tooth in their heads, but no one doubts their efficiency as predators and scavengers. A sharp beak can as useful as teeth, and at least it avoids a lot of dental problems in later life.

## ● Scenes of the hunt

It would seem obvious that any attempt to figure out how the predatory dinosaurs hunted must be entirely guesswork. Well, not entirely. A set of fossil tracks in Australia shows masses of small dinosaurs trotting about. Over to one side is a track made by a large *Allosaurus*-like predator tearing on to the scene. In front of it, the small footprints scatter: you can almost see the little dinosaurs rushing off in confusion.

## ● Pack hunting

John Ostrom speculated that *Deinonychus*, the little flesh-eater he found in 1964, might have been a social pack-hunting animal. He drew a comparison with modern hunting dogs, which operate as a pack, communicating with each other, in order to bring down a prey animal, like a wildebeest, that is much bigger than themselves. Ostrom speculated that *Deinonychus* did the same, because the only suitable prey animal found in the same beds was the large ornithopod *Tenontosaurus*. Perhaps, he suggested, half a dozen *Deinonychus* would gather round one *Tenontosaurus* and harry it, leaping and slashing. Eventually the large herbivore might keel over, weakened by loss of blood, and the predators would move in. This is in contrast to the solitary hunting style of modern cats, such as lions and tigers. Evidence came to light with the discovery of a *Tenontosaurus* kill site: a skeleton of the herbivore was surrounded by remains of four individuals of *Deinonychus*, plus dozens of *Deinonychus* teeth, perhaps broken off during the tussle.

## ● Fight to the death

An astonishing specimen of a dinosaur fight came to light in 1970. A Polish-Mongolian field crew found a skeleton of *Velociraptor*, a fiendish predator closely related to *Deinonychus*, locked in mortal combat with a *Protoceratops*, a modest-sized ceratopsian dinosaur, a herbivore. The *Velociraptor* has got one of its powerful hands hooked round the ceratopsian's head shield, and its vicious toe claws are raking at the herbivore's belly. Why did they die together? Perhaps *Protoceratops* had managed to head-butt the *Velociraptor* or wound it seriously with its thickened snout (*Protoceratops* didn't quite have a nose horn, as seen in later descendants, such as *Triceratops*). Anyway, both seem to have died in each other's arms, perhaps weakened by exhaustion and blood loss.

## ● Bite marks

Primary evidence for the diet of carnivorous dinosaurs has come to light recently. Several bones of herbivorous dinosaurs from the Late Cretaceous of North America were found with large puncture marks. Paleontologists poured in some casting rubber compound, let it set and pulled it out. They had a perfect four-inch-long *Tyrannosaurus* tooth. So there is no doubt that *Tyrannosaurus* bit into the flesh of contemporary herbivores, such as *Triceratops* and *Edmontosaurus*, with such force that it drove its teeth into the bone up to the gum line.

## ● Dinosaur cannibals

Surely dinosaurs were indiscriminate diners. They were obviously sly, evil-minded reptiles, so dads wouldn't have quibbled about eating their own babies if they were hungry. This seems to be true, at least sometimes, and it's not clear that moms didn't also eat baby dinosaurs. A famous death assemblage of dinosaurs, hundreds of skeletons of the modest-sized early theropod *Coelophysis* from the Late Triassic of Arizona, showed cannibalism in practice. One of the adult skeletons, excavated in 1947 from the appropriately named Ghost Ranch site, had a neatly curled-up baby in its stomach. Was this a mistake, the result of extraordinary famine, or normal practice?

## ● Did dinosaurs urinate?

A controversial image in the first *Walking with Dinosaurs* program shows *Postosuchus* urinating copiously. Two adults have just had a fight for territorial dominance, and the victor urinates to establish its presence. This is classic mammalian behavior. Of course, *Postosuchus* is not a dinosaur but a basal archosaur, close to the line that led to crocodiles. But the logic is similar for *Postosuchus* and dinosaurs. Modern crocodiles and birds, closest living descendants of the dinosaurs, do not produce copious quantities of liquid urine. In fact, they get rid of nitrogenous waste from their bodies in crystalline form (uric acid): the uric acid is the white bit in a bird dropping. Mammals, and various aquatic groups, dissolve their nitrogenous waste in water and get rid of it in the form of urea. So, *Postosuchus* and the dinosaurs might just have urinated in this luxurious manner, but it's more likely that they did not.

CHAPTER FIVE

# WALKING AND RUNNING

Dinosaurs obviously walked, ran, galloped, crawled, and hopped about. But how did they move? How fast could they go? How sure can we be? Could *Apatosaurus* gallop as fast as an express train? Did *Plateosaurus* run on its hind legs or on all fours? It would be easy to think that you'd need a time machine to answer these questions. But paleontologists can actually use three or four different ways to get at the answer. Often, surprisingly, there are several strong independent lines of evidence that can confirm exactly how a particular dinosaur moved about.

## ● Comparison with modern animals

As we've seen in Chapter 1, a key principle in paleobiology (the study of how ancient plants and animals lived) is actualism. That's a fancy word for looking at what happens today, and applying the principles to the past. Actualism is the claim that the laws of physics, and of nature, haven't changed. If water freezes at zero degrees centigrade today, so it did also in the Mesozoic. If a heavy animal like an elephant has legs like tree trunks today, so, too, would a heavy animal in the past.

## ● Runners and jumpers: it's all in the leg shape

You can tell if someone is a trained athlete by looking at his or her body shape. And it's the same – but more so – with different species of animals. Jumpers, like rabbits and kangaroos, have a very special kind of back leg. Climbers, like monkeys or squirrels, have specialized grasping hands and flexible shoulder joints. Slow-movers, like bears and humans, put their feet flat on the ground (remember, that's called a plantigrade foot; see Chapter 1). Fast runners, like dogs, horses, and deer, have specially elongated legs, and they are up on their toes (the digitigrade stance). No matter that some, like cows, don't run very fast now: their undomesticated relatives did.

## ● Bone: more alive than dead

Bone is both living and inert (i.e. non-living). It is both strong and bendy, hard and soft. It is a composite material, part living, part not living. The living material is a flexible protein called collagen, and blood vessels and fat bodies within the bone keep it in good shape. The other half is crystalline apatite, calcium phosphate, deposited as tiny needles on to the collagen. So, a living bone can bend, but only so far. A dead bone, a bone that has been removed from an animal remains tough and bendy for a few months after death. So a butcher's bone still retains collagen. But after that, the proteins and other living materials decompose, and you're left with the inert, crystalline part. This inert part snaps like a dead twig.

## ● Bone strength

Bones can break. One of the key kinds of actualistic observations in biomechanics (the study of how plants and animals work) is to consider the strengths of biological materials. Bones are pretty tough. They have to be, since they form the framework, the skeleton, that holds the body up. Importantly, the bones provide the anchoring points for all the muscles. Without your skeleton, you'd be a floppy jellyfish-like heap that couldn't move. Biologists can test the breaking strength of bones in simple machines: you keep adding weights until the bone snaps. The obvious actualistic assumption is that dinosaur bone strengths were the same as the bone strengths of modern animals.

## ● Speed and impact

It's important to think about what happens when you start to move. This is critical for understanding how fast an animal can go, whether it's a living animal or an extinct one. When you are standing still, your weight can be considered as a single unit of mass. At rest, standing up, half your mass is supported by each leg. When you begin to walk and lift a foot, immediately the stress on your leg doubles: one leg lifted means all your mass is pressing down the leg that's still on the ground. So your ground force, as it's called, has doubled. But when you're walking, your foot hits the ground with a little bang. It's not much, but it's enough to bring the ground force up to 2.5 times normal. When an athlete sprints, his or her ground force increases to 3.6 times normal. When he or she jumps, the figure rises to 5 or 6 times normal. The legs have to be able to take this extra battering.

## ● Safety factors

A basic principle of biomechanics is the safety factor. This is the built-in allowance for unusual extra forces that might happen very rarely. Clearly, human legs have a safety factor of at least five or six times normal ground forces. You have to jump from a pretty unusual height to break your legs. Every part of your body is engineered with a safety factor of about ten times. But as animals get bigger, the safety factor reduces. In other words, we're more safely engineered than a cow, a cow more so than an elephant. An elephant has a safety factor of only two or three times at most, so it has to be careful about jumping off walls!

## ● Bone width and body weight

Skeletons are finely honed for their purpose. But the safety factor diminishes as the size of an animal increases. This is clearly an important fact to remember when we look at dinosaurs. A final key biomechanical principle to understand is that the fatness of a leg depends on the weight of an animal, not its area in side view. In other words, leg width, a two-dimensional measure, actually depends on body weight, a three-dimensional measure.

## ● Elephants scaled like gazelles?

The area v. volume effect is very obvious if you look at silhouettes of a gazelle, a horse, and an elephant. If you simply scaled up the gazelle,

**A sauropod dinosaur compared with the biggest living land mammal, an elephant. The diameter of the leg bones (shaded black), depends on body weight, not body length or height.**

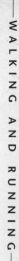
with its matchstick-like legs, to elephant size, it would be ridiculous. Its legs would break immediately. That's because the elephant is a hundred times the weight of the gazelle, but its silhouette area is perhaps only ten or twenty times larger. This calculation will allow us to tackle, in a moment, the question of just how big a land animal can be. Scaling up further, comparing an elephant and a sauropod dinosaur (see page 78), the effect is clear.

## ● The galloping *Brachiosaurus*

Could Bob Bakker's image of a galloping *Brachiosaurus* (see Chapter 2) really be true? Of course not. Think about it. If an elephant has a built-in safety factor of two or three, *Brachiosaurus* had an even tighter allowance for the unusual. If it jumped off a three-foot-high wall, it would collapse in a heap of broken bones. Galloping was clearly out. Probably the best it could do was a leisurely amble.

## ● Back to first principles

How far can paleobiologists take their biomechanical reasoning? Remember, the technique doesn't allow you to say with hundred percent certainty that *Brachiosaurus* could not gallop. But I think you'll agree that this is 99 percent certain: it's so much a matter of common sense, and the simple application of biomechanical rules that it would take a miracle in the design of *Brachiosaurus* to get around it. Now, with powerful computers, it should be possible to write equations and make virtual moving models that can test different modes of locomotion for these animals.

## ● Weighing dinosaurs

The first step in a biomechanical study is to weigh your dinosaur. Impossible? Well, yes and no. Without a time machine, a huge pair of weighing scales and a great deal of mad stupidity, we will never know how heavy *Tyrannosaurus* or *Brachiosaurus* were. But again the basic rules of nature come into play. Animals with small skeletons weigh less than animals with large skeletons. A dinosaur with a skeleton the same size as a cow weighed the same as a cow. Reasonable? So you can make reasonable guesses just from the size of the skeleton. *Brachiosaurus* is twenty times the size of an elephant, so it weighed 3.3 tons (3 tonnes) x 20, which is 66 tons (60 tonnes). Not a bad estimate.

# ● Archimedes' streak

In 1962 Ned Colbert, a famous North American vertebrate paleontologist, thought of a clever trick to get a better estimate. He remembered Archimedes' principle, which I'm sure you do, too. Archimedes was the man who, so they say, jumped into a bath full of water. A great slosh of water jumped out of the bath as he jumped in. And he ran off down the street, stark naked, shouting, "Eureka", "I have it." The volume of water he displaced when he jumped into the bath was exactly equal to his body volume. If you know how much a quart, or a gallon, or whatever, of flesh weighs, then you can figure out the weight directly from the volume. History does not record whether Archimedes cleaned up the mess on the bathroom floor, nor what Mrs. Archimedes had to say about it.

# ● Dangling plastic models in a beaker

Ned Colbert realized that he could dangle plastic models of dinosaurs in beakers containing exactly one litre of water. Of course, the trick was to have a high-quality scale model that was carefully sculpted and based on studies of actual specimens. The volume that was displaced equalled the volume of the model. Knowing the scale of the model, Colbert was able to multiply the volume up to that for a full-sized dinosaur. He then found, by cutting out cubes of dead crocodiles, that a one litre volume of crocodile flesh pretty much equals one kilogramme of weight. (The metric measures were designed this way so that one litre of water weighs exactly one kilogramme, and since animals are eighty-five per-cent water there's no problem.)

# ● Debates about weight

In 1962 Ned Colbert estimated some pretty acceptable weights for well-known dinosaurs, and these have been quoted all over the place. For example, Colbert calculated that *Brachiosaurus* weighed 85 tons, *Tyrannosaurus*, eight tons. His figures were debated, of course. (Paleontologists, like all scientists, are very argumentative. Remember the definition of a scientist: a scientist is a child who won't stop asking questions.) So Professor McNeill Alexander, a British biomechanics expert, recalculated *Brachiosaurus* at 51 tons, and Greg Paul, an American dinosaur enthusiast and artist, calculated a piddling 35 tons. Why the disputes? It was to do with the plastic models they each used. Colbert had a fat one, Paul had a thin one. So there are some problems with this technique. Maybe it can give weights only to within a tolerance of plus or minus fifty percent.

## ● Smallest

Which was the smallest dinosaur of all time? The smallest is usually believed to be *Compsognathus*, a slender flesh-eater from the Late Jurassic of southern Germany. In fact, it comes from the same quarries as *Archaeopteryx* (see Chapters 3 and 9). *Compsognathus* was only two feet long, most of which was a whippy tail. It stood no taller than a turkey, and probably weighed a mere 22 pounds. Even smaller were some juvenile and embryo dinosaurs. Now there are even embryos and hatchlings, some of them no more than ten inches long, and probably weighing two pounds or less.

## ● Largest

A very popular question is which was the largest dinosaur. Well, the biggest dinosaur was certainly a sauropod, and the biggest reasonably well-known dinosaurs are still *Brachiosaurus* (75 feet long; 42 feet tall; 55 tons in weight) and *Diplodocus* (90 feet long; 22 tons in weight). Several record-breakers were announced in the 1970s and 1980s, all sauropods, but with ever more reckless names: *Supersaurus*, *Ultrasaurus*, and *Seismosaurus*, all from the Morrison Formation (see Chapter 4). Early claims gave some of these body lengths of 300 feet and weights of over a hundred tons. But none of these is yet known from a complete skeleton, and size estimates have been cut back. But we still have the 100 foot-long/ 77 to 100 ton sauropod to contend with.

## ● Is there a size limit?

Speculating about the weights of known sauropods leads to an interesting question: is there a maximum size limit? Surely, beyond a certain size the whole skeleton would collapse? Well, it's not collapsing skeletons dinosaurs have to worry about, but locking legs.
Remember the area vs. volume problem. As animals get bigger, their legs increase in width in proportion to weight. Elephants have legs like tree trunks, and gazelles have thin, spindly legs. Sauropods had even more tree trunk-like legs than elephants. As the body weight of the hypothetical giant sauropod increased, leg diameter increased at the same rate. It's been calculated that at about 165 tons, a giant sauropod would come to a complete standstill: its four legs would be

so wide that they would meet in the middle, and it wouldn't be able to walk. So there you are.

## ● Chopping a dinosaur into segments

Playing with plastic models is one way to find out the weight of a dinosaur. Is there another way? Donald Henderson, now at Johns Hopkins University figured out a way to study dinosaurian biomechanics from first principles. The first step was to estimate body weight. He cut up a dinosaur's body into imaginary segments. The volume, and hence the weight, of each segment could be calculated easily by computer. The key for this method, published in 1999, was that each segment could be adjusted for local conditions. For example, using plastic models ignored the fact that densities vary in different parts of an animal. The lungs are especially important. Just as in humans, dinosaurs had a pair of lungs, and these were empty spaces in the chest region which effectively weighed nothing.

## ● Finding the center of balance

The center of balance is the exact center of the animal, and it's an important thing to know for any study of locomotion. Having calculated the weight of each segment of his dinosaur, Donald Henderson identified the exact location of the center of mass (or center of balance). You can find the center of mass of a model by dangling it on a string. When you've got the string attached at the right spot, the model dangles in a perfectly balanced way. In the theropods, the center of mass lay just in front of the hips, squarely above the feet, just as it should. So far, so good.

## ● Wiggling your hips

When you walk, your hips swing from side to side, in some people more so than in others. This is not simply a method to attract members of the opposite sex. It is an essential biomechanical operation. As you begin a stride, say, lifting your right foot, your whole weight swings over the left leg. So, right leg up, and hips swing to the left; left leg up, and hips swing to the right. If you didn't do this, you would fall over. Try it.

## ● Nodding your head

During walking, the body has to respond to the risk of falling over in three dimensions. Side-to-side swaying of the hips only describes one axis. Look at a person walking in side view. As he takes a stride, moving the right leg forward, the body and head tip back a bit. As the right foot hits the ground, the body leans forward over that foot. Up comes the left foot, back goes the body, then forward over the left foot as it touches down. You see this effect in an exaggerated way in birds. Just watch a pigeon or a peacock trotting about. The legs move in a fast jerk, and the head nods back and forward furiously. If you tied a pigeon's neck to a rigid splint so it couldn't nod, it couldn't walk – not something to try.

## ● Looking into joints

Once the center of mass has been established for a dinosaur, it is critical then to consider the bones themselves. You can actually manipulate the fossil bones in your hand, and find out exactly what kinds of joints they had. The shape of the ends of the bones is governed precisely by the kinds of movements they made in life.

## ● Ball-and-socket and hinge joints

The main kinds of joints between bones are ball-and-socket, where the limb can rotate and twist round a wide arc, and hinge, where it simply flaps back and forward or up and down. Our hip and shoulder joints are ball-and-socket type. You can swing your arm through 360 degrees (as noted in Chapter 1, a throwback from our tree dangling ape heritage). You can't swing your leg around quite so much, unless you have unusual abilities, but it's a ball-and-socket joint again. Human knees and ankles, though, are pretty much simple hinges. You can flap your leg back and forward with these joints, but unlike ball-and-socket joints, there's not much possibility of sideways movement.

## ● Joint limits

Bones are designed so that the joints will not dislocate. You may know how horribly painful it is if you dislocate your shoulder or elbow. Normally, the joints are held together tightly by tendons over the joint, by a fluid-filled tough balloonlike structure called a bursa and by the shapes of the bones themselves. There are special processes, or extensions, on the bones at your knee and elbow joints that stop them swinging over and bending the wrong way. It's exactly the same for all other animals, living and extinct. Well-preserved dinosaur bones show perfectly which ways they could bend, and by how much.

## ● Muscles only pull

A key thing to know is that muscles only pull; they never push. Muscles are made from millions of fine fibers arranged in bunches. They pull by contracting: the fibers slide over each other and tighten up. But a muscle cannot push by forcing itself to be longer. Pushing is done by having opposing muscles on the back and front of each limb. So you tighten your arm (pull it forward) with the biceps, and you pull it back by the triceps, which is attached along the back of the shoulder and arm. The final point is that muscles can contract only by a predictable amount, maybe twenty to thirty percent of their length. All this is important in reconstructing the locomotion of ancient animals.

## ● Muscles in ancient animals

Is it just guesswork where you put the muscles on a dinosaur? Not so, as we've seen in Chapters 1 and 2. Dinosaurs almost certainly had the same muscles as birds, crocodiles, and indeed all other living tetrapods. So we know exactly which muscles any dinosaur had around its legs and arms. The sizes of the muscles can be estimated pretty accurately from the sizes of the bones: thick bones equal thick muscles, on the whole. Also, the bones themselves speak to the paleontologist.

## ● Muscle insertions

Muscles are attached to the bones by long fibers that run into the structure of the bone; they aren't just stuck on the surface. If they were just stuck on, body builders would have real problems: they would go for a lift, tighten their biceps and then pull themselves to pieces! So in a well-preserved dinosaur bone you can quite easily see these fiber insertion areas as roughened patches. The area of the insertion patches is an exact guide to the position and size of the ancient muscle. There's not much guesswork in reconstructing dinosaur musculature.

## ● Scanning the bones

In figuring out how a particular dinosaur moved using computer-based calculations, the first step is to scan in the bones. From the bones themselves, and from carefully measured drawings, Donald Henderson input the complex 3-D shape of each of the bones of the legs and hips. At least, when he had scanned in a left femur (thighbone), he just flipped it in the computer's memory to generate the right femur. It takes a few days to scan all the bones to the level of detail required.

## ● Coding the joint movements

The second step in the calculation was to code in the joint movements. Bone manipulations and measurements show exactly how each joint worked, and how far the limb could bend, and in which directions. All of that was written into the program. The legs were complete, and Henderson could manipulate them on the screen to see the maximum bend forward and backward in a stride. But that still wasn't enough.

## ● Calculating the step cycle

The critical stage in calculating the locomotion pattern of a dinosaur is the hardest part. This was to write a series of equations, one for each joint in the legs, and then set the computer to solving the equations simultaneously. This took huge computing power, and it had to be done on a mainframe computer at the University of Bristol. A desktop machine couldn't cope, or it would take a week to complete each set of calculations. As the mainframe whirred into action, the dinosaur legs were shown walking slowly across the screen: a real-time biomechanical calculation that was rendered as a cartoon. In single frames, a whole stride cycle could be shown.( To see Henderson's moving images of dinosaurs walking go to web site http://palaeo.gly.bris.ac.uk/).

## ● What *Tyrannosaurus* could and could not do

So could *Tyrannosaurus* run as fast as a racehorse, in hot pursuit of a small, fleet-footed herbivore? Or was its best effort a leisurely stroll, like a schoolboy on his way to the principal's office? This had been an unresolved debate for twenty years. Henderson found that it was something in between. Faster than 30 feet per second, and *Tyrannosaurus* would have collapsed in a heap. The system failed because his legs would slide away from under him, and his body could not keep up the necessary sideways and back-and-forward movements to maintain his center of mass directly over his feet. Henderson's work lay behind the striding of *Allosaurus* and *Tyrannosaurus* in *Walking with Dinosaurs*. If you thought the moving images were just a cartoonist's clever trickery, think again! Three years of intensely complex biomechanical calculations were in there as well.

## ● Fossilized behavior

You don't have to believe the complex biomechanical calculations just because I say so. There is an entirely independent test of whether the postulated leg movements are correct or not: dinosaur footprints. The ancient tracks of animals have been called "fossilized behavior" by Dolf Seilacher of Tübingen (Germany) and Yale (Connecticut), a famous expert on trace fossils (i.e. burrows, tracks, and trails). He was right. When you look at a track of dinosaur footprints, you are looking at the shape of the flesh of its foot, and you are looking at a real sequence of events 150 million years ago.

## ● How footprints are preserved

There are millions of dinosaur footprints, most of them exposed to the elements in the quarries where they were found. Some tracks have been taken to museums, but they're so huge that it's often best to see them *in situ*, on the ground. All we know is that dinosaurs, and many other animals, ran about, burrowed, laid eggs, fed, and yes, even copulated on the mud or sand. Usually their prints would be lost by water or wind erosion. But sometimes they are preserved, especially if the sand or mud was wet but not too sloppy. Dry sand might be blown, or washed, over, and the footprints are fixed for ever.

## ● Stride and pace

In a run of dinosaur footprints you can see many things. You can measure the size of an individual footprint. You can also measure the pace and stride length by laying a ruler down between repeats of the same foot, and the two feet (see page 87). A pace is the distance from a left to a right footprint. A stride is the distance between the impressions of the same foot. So if your pace is three feet, your stride is six feet. The pace angle can also be figured out: this is the angle from a right to a left or a left to a right imprint.

## ● Identifying the trackmaker

The size of a dinosaur footprint gives some clues to its maker. Also, the shape of the foot and the number of toes can be seen easily. Most bipedal dinosaurs made three-toed prints (see page 87) while sauropods made big stumplike prints. Medium-sized quadrupedal dinosaurs, such as prosauropods, ceratopsians, ankylosaurs, and stegosaurs, made partially stumplike, partially separate-toed prints. In more detail, a well-preserved footprint can even show the separate joints of each toe, and paleontologists can fit foot skeletons to prints, and sometime even pin down which species of dinosaur made a particular track.

**Dinosaur tracks. LEFT: Measuring stride length (dashed line), pace length (solid line)and pace angle. RIGHT: An ornithopod footprint with matching foot skeleton superimposed.**

## ● *Tyrannosaurus* tracks

*Tyrannosaurus* tracks give an independent test of Donald Henderson's computer calculations. First, the tracks show that the right and left feet fell almost perfectly along a single line. So as the animal lifted its right foot, it swung over and the left foot came down bang in the midline. As it lifted its left foot, the body swung the other way, and the right foot came down precisely in front of the last left footprint. Henderson matched his calculated *Tyrannosaurus* walking patterns, and he found that they gave exactly the same imaginary track as in the real examples. First test passed. He also ran a blind trial for running patterns of an ostrich. Again, the computer churned out exactly the right measurements when he compared it with a real ostrich running. Second test passed. Method works.

## ● Calculating speeds

Many people have noticed that as an animal moves faster, the pace and stride lengths increase. This is no surprise. Try walking, and then running, along a wet beach. If your walking pace is three feet, your running pace may be six or more feet, and your footprints preserve a permanent record behind you. Pace and stride length increase in proportion to speed. This raises interesting possibilities.

## ● Professor Alexander's useful formula

Professor McNeill Alexander noticed the speed and stride length effect, and in 1975 he had the insight to make it work for dinosaurs. He established a simple equation (pay attention in the back there):

Speed = Square root of (leg length x gravity)/Froude number

Speed is obvious. Leg length is the length of the leg of the running beast, measured from the hip joint to the ground. Gravity is the normal gravitational acceleration, which is essentially 30 feet per second per second. Don't worry about it; just remember the 30. And then there's the Froude number.

# The Froude number

Alexander's insight into calculating dinosaur speeds was a bit of lateral thinking. The Froude number is a physical constant that shipbuilders use to calculate how well a ship will cut through water. The Froude number varies with the size and shape of the ship, and it's a way of calculating the ideal streamlined shape for most efficient movement. Alexander realized that the Froude number might be useful for calculating animal running speeds. He tested his formula on dozens of modern animals, from hamsters to rhinos, and they all fit its predictions precisely, whether they run on two legs or four.

# The fastest dinosaur?

Which was the fastest dinosaur? Paleontologists scoured the world for dinosaur tracks and began calculating speeds using Alexander's formula. Most of the figures came out in the range of six to thirty feet per second. Fastest of all were the small long-legged theropods, such as the ostrich dinosaurs *Struthiomimus* and *Ornithomimus*. They could manage an amazing 65 feet per second, as fast as a racehorse, or a motorist speeding in a city. But such speeds were probably just for short dashes to safety, or chasing after a tasty lizard or beetle.

# Biped vs. quadruped

Dinosaurs walked on two legs (bipeds) or four (quadrupeds). There are advantages with each walking style. As bipeds ourselves, we can see how it frees the arms and hands for other functions, such as grabbing and carrying things. It also allows us to be taller, so we can look around for danger more easily. You feel pretty defenceless if you go down on all fours: try it in your local supermarket. Dinosaurs were half-bipedal, half-quadrupedal. How did this happen? Once, it was assumed that quadrupedality had come first in evolution since all other reptiles are quadrupeds. However, we now know that the first dinosaurs were smallish bipeds (see Chapter 9). Quadrupeds evolved from these bipeds, and they usually went back down on all fours when they became too big to carry on upright.

## ● Reptiles erect and sprawling

Not only were many dinosaurs unusual for reptiles in being bipedal, but they also all stood erect. If you look at a typical lizard or a crocodile, its legs stick out to the sides. The upper part of the arm and leg is pretty much horizontal, and the calf and forearm are vertical. When they walk, the arms and legs sweep out to the sides, as well as going back and forward. Dinosaurs were fully erect. Their arms and legs were tucked straight under the body, just as in mammals and in birds. The upper part of the leg and arm was vertical, and essentially the whole length of the leg made the stride. So erect posture in the first smallish dinosaurs may have given them a longer stride than their enemies. Erect posture certainly allowed dinosaurs to do lots of things.

## ● Later advantages of erectness

Erect animals can be bipedal and they can be big. It's virtually impossible to see how a lizard could run bipedally in any serious way, or indeed how a lizard could become seriously large. Erect dinosaurs were also able to free up their arms for other purposes. One of those purposes was flight: the arms became wings in birds (see Chapter 9). No sprawler could do that. So are these the reasons why the first dinosaurs stood erect: so they could be bipedal, large, and winged? No, of course not. Evolution works only at the present; it doesn't make predictions. The first dinosaurs were erect since it helped them to run fast and escape being eaten. That's all.

## ● Megatracksites

Dinosaurs were superb long-distance walkers. Some tracksites extend over hundreds of yards, even miles. Paleontologists have followed some ancient river banks over huge distances, and they find tracks of herds of dinosaurs trotting along beside the water. These so-called megatracksites are especially common in the Late Jurassic and Early Cretaceous of the midwestern states of North America. But what were these dinosaurs up to?

## ● Mass migration routes

It seems likely that some dinosaur megatracksites represent parts of long distance migration routes. Some of them, for example, are found all along the western shore of the great Mid-American Seaway, running from Alaska in the north to Texas and Mexico in the south. Perhaps these amazing tracksites record herds of dinosaurs trekking thousands of miles in search of food or warmth. They may have headed north in summer to find cooler weather and lusher vegetation. As winter came on, they headed south, along well-worn trails, staying in a pleasantly warm climate belt. Paleobiologists look at modern elephants and moose that cover huge distances in search of food when they have cleared out one area. A herd of voracious sauropods, each eating tons of food per day (see Chapters 4 and 6) would soon clear out all the plants in an area like the modern elephans and moose.

## ● Dinoturbation

Dinosaurs had a major effect on their environments. It's well known that elephants create that typically African landscape of lush grass with rare low trees scattered here and there. Fullscale forests can never become established because the elephants eat the young trees and trample them. Dinosaurs were so common in some places that their footprints churned the lakes and rivers, crushing other animals as they went. Fossil tracks record a sediment-modifying process called "dinoturbation." This grand word says it all: it's the effect of tons of dinosaur flesh passing a single spot and thoroughly churning the sand or mud into a structureless mush.

## ● Beware of time-averaging

Dinosaur track experts always warn us of one thing. We must beware time averaging. This is a classic phenomenon seen in all fossil deposits. One thin band of rock may be absolutely stuffed with shells, or bones, or footprints, but we mustn't assume that they were dumped there in an instant. For the shells or bones, the thin layer might represent years of slow accumulation, with a few shells or bones being washed in each year. A single rock surface covered with footprints might be showing us ten years of accumulated activity. A huge mass of footprints could be made by a lone dinosaur trotting over a mud patch once a day to a waterhole. You can detect time averaging by checking whether all the footprints are preserved to exactly the same quality, or whether some have been worn away by exposure to the elements and others have not.

## ● Swimming dinosaurs

A classic dinosaur scene shows a herbivore chased by a pack of carnivores. The herbivore plunges into a lake and swims off, leaving the carnivores on the shore roaring and screaming with frustration. We know that the herbivores could swim. There is a wonderful track of an *Apatosaurus* in the Morrison Formation. It's walking along through lakeshore mud and then plunges into the water. Instead of all four feet touching the bottom, you just see the odd handprint, as it steered itself by jabbing into the lake bed. (Or it could be a sauropod walking on its hands, legs up in the air. Not likely.) But stop a minute. This is complete nonsense! Why would the theropods stop dead on the shore?

## ● Theropods can swim too

The myth of non-swimming predators probably comes from observations of cats. Most cats don't much like to get wet, and they'll do anything to avoid a bath. But in the wild, cats swim. Also, think of dogs: you can't keep them out of the water. So theropods could probably swim as well. This was proved in the 1970s with the discovery of a track produced by a theropod which was paddling itself along in the water with a leisurely kick into the bottom mud from time to time.

WALKING AND RUNNING

## CHAPTER SIX

# WARM-BLOODED DINOSAURS?

A fascinating long-running debate about dinosaurs has been whether they were warm-blooded or not. In itself, this might seem a bit arcane, but warm- vs. cold-bloodedness has wide ramifications. It concerns the whole biology of an animal. Warm-blooded animals, like birds and mammals, are active day and night, and generally live life at a high rate of activity. Cold-blooded animals, like most modern reptiles and amphibians, have a slower pace. The extreme image is the large snake or lizard which eats once a week and then seems to snooze most of the time while it digests its meal. The solution to this debate, then, could determine how we view dinosaurs: as active, fast-moving, exciting, dynamic beasts, or as sluggish, slow-moving, unintelligent automata.

## ● Professor Owen's speculation

In 1842 when Richard Owen coined the name "Dinosauria," he speculated that these giant reptiles had been rather mammal-like in their physiology; very different from living reptiles, such as lizards and crocodiles. Owen clearly thought that dinosaurs were able to control their body temperature to some extent, and to keep it high. But remember that he had a special reason for such speculations (see Chapter 2): he believed deep down that dinosaurs were much more advanced than modern reptiles are.

## ● But dinosaurs are reptiles

Most Victorian paleontologists didn't really like the notion that the dinosaurs were warm-blooded. They were too influenced by the obvious fact that (a) dinosaurs were reptiles, and (b) reptiles are cold and slimy. Enough said. Thomas Henry Huxley, who clearly argued that birds evolved from theropod dinosaurs (see Chapter 3), could have rocked the boat. After all, birds are warm-blooded, and if birds are living dinosaurs, then perhaps their ancestors could have been warm-blooded also. Huxley didn't go through this argument seriously, perhaps because he didn't want to be seen to be supporting Owen in any way. Huxley thought Owen was an old buffoon, and Owen thought Huxley was a young nincompoop. Such is life.

## ● Hair in pterosaurs

In about 1920 some German paleontologists spotted what they thought were fossilized hairs on the skin of pterosaurs, the flying reptiles. These were pterosaur fossils from the same localities in southern Germany that had produced the *Archaeopteryx* skeletons. The feathers of *Archaeopteryx* were preserved, and so were the pterosaur hairs. The particular southern German rocks were so-called lithographic limestones, very fine limestones that were used as printing plates. They were of huge interest to paleontologists, though, because the fossils were all exquisitely well preserved. You could find worms, insect wings, lizards with skin, birds with feathers, and pterosaurs with hair intact. Fantastic. But what about those hairy pterosaurs?

## ● Hairs, feathers, and insulation

Hair is an insulating medium, and any hairy animal must then be warm-blooded. Among modern vertebrates, mammals have hair, and birds have feathers. Both groups use these soft, airy skin coverings to create a zone of trapped air around their bodies which keeps the heat in. A fur coat really. Reptiles, amphibians, and fish don't have an insulating covering, but they're cold blooded, so it doesn't matter. If pterosaurs had hair, that means they were warm-blooded. So perhaps their close relatives, the dinosaurs, might also have been warm-blooded.

## ● The lizardly consensus

Dinosaur paleontologists ignored the suggestion that dinosaurs might have been warm-blooded for most of the twentieth century. They stuck firmly to the Victorian view. Dinosaurs are overblown lizards. Lizards are cold and slimy and sluggish. So dinosaurs were cold and slimy and sluggish as well.

## A time machine and a rectal thermometer?

Gary Larson has an excellent cartoon of an intrepid, if rather wild-eyed, explorer, leaving his time machine and cautiously approaching the rear end of a dinosaur. The professor is carrying a huge thermometer. The caption reads,'An instant later, both Professor Waxman and his time machine are obliterated, leaving the cold-blooded/warm-blooded dinosaur debate still unresolved.' So maybe we will never know. Can we ever assess the body temperature of an extinct animal? Of course not. But by using the principles of actualism, the assumption that the laws of physics and of nature have not changed since the Mesozoic, we can begin to debate the issue.

## Dr Bakker's wild proposal

In 1970 Bob Bakker, while still a graduate student at Yale University, made an astonishing, shocking, stimulating proposal. He suggested that the dinosaurs had not been cold-blooded and sluggish, but that they had been warm-blooded, just like birds and mammals. His new vision conjured up images of dinosaurs leaping, cavorting, jumping and hurtling around like racehorses. He marshaled a range of evidence that the dinosaurs had been endothermic homeotherms. Now, there's a mouthful! Endothermic homeotherms. To understand this we must get a grip on the physiology of modern vertebrates.

## What it means to be warm-blooded

Modern animals divide up into two main categories in terms of temperature control. The ectotherms, like fish and reptiles, generally use only external means to control their body temperature. For example, lizards bask on rocks to raise their temperatures, and hide in holes to cool down. The sea at depth never freezes, so marine fish are generally adapted to live at temperatures somewhere above freezing. The endotherms, like birds and mammals, can maintain warm body temperatures by internal means, by burning up food and by a complex feedback mechanism that heats and cools the body to maintain temperature at a precise level. There's another way to look at this.

## Look at it another way

Modern animals can be divided into poikilotherms and homeotherms. Poikilotherms have variable body temperatures, and homeotherms have constant body temperatures. Lizards are clearly poikilothermic

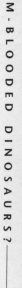

ectotherms, and birds and mammals are generally endothermic homeotherms. But poikilotherm does not always equal ectotherm, or homeotherm equal endotherm. Confused?

## ● Now, let's revise that

It's necessary to use four terms to describe the thermal physiology of animals because you can find all possible combinations in nature. For example, fish are ectothermic homeotherms: their body temperature is constant, although controlled externally, since the temperature of the sea does not change much. Bats and hummingbirds are poikilothermic endotherms, since they can switch off their expensive heating system at night, or in winter. Admittedly, most animals are either ectothermic poikilotherms or endothermic homeotherms. You have to keep in mind the exceptions. But are these categories absolutes?

## ● A sliding scale?

Temperature control categories of living animals aren't fixed absolutes. It's a common belief that animals are either one hundred percent endotherms or one hundred percent ectotherms. Reptiles are cold-blooded, mammals are warm-blooded, and that's that. Not always. When you look in detail, you can find intermediates. Most mammals hold their body temperatures quite high, at about 98.6°F (37°C), just as humans do. But some mammals, such as the duck-billed platypus and the sloths, hold their body temperatures much lower, at 82°–86 °F (28–30°C). Also, tunas are fish, but they maintain high active body temperatures. So, too, do some of the large reptiles, such as sea turtles, large crocodiles, and snakes. They are ectotherms, but they can somehow switch on high temperatures when they need to.

## ● But is endothermy always best?

Surely warm-bloodedness is best? We're warm-blooded, and mammals and birds are obviously much more successful than reptiles. To an extent this is true. Endothermy certainly allows birds and mammals to live in cold regions where reptiles would be torpid. Endothermy, by maintaining a constant internal temperature, also has an advantage in allowing biochemical reactions to occur at maximum efficiency at all times. With variable body temperatures, these reactions run slow and fast. But is endothermy really always best?

## ● Sometimes it's good to be an ectotherm

Ectotherms have huge advantages. They can withstand wide
fluctuations in body temperature, sometimes by as much as (20°C) each
day. A change of only a few degrees can be critical for an endotherm:
you feel pretty rotten and your doctor tells you that you are feverish if
your body temperature goes up by just one degree.

## ● Endotherms are what they eat

The biggest disadvantage for endotherms is the amount of food and
water they need. On average, an endotherm has to eat ten times – yes
ten times – as much as an ectotherm of the same weight. That's why
snakes and crocodiles seem to laze around so much. While you are out
hunting for enough food to keep your internal fires stoked up, the
crocodile just blinks sleepily and tut-tuts wisely. Endotherms also burn
up ridiculous amounts of water. We're forever drinking. Many lizards
and snakes can get by on almost no water at all, and they can live in
deserts where no endotherm could survive. So which is best?

## ● Bakker's motivation

In his 1970 paper about dinosaur warm-bloodedness, Bob Bakker
argued that the dinosaurs must have been endothermic because they
were hugely successful. In other words, he was making a simplistic link
between thermal state and overall success. Endothermy is better than
ectothermy, dinosaurs are successful, so dinosaurs were endothermic.
He marshaled five main lines of evidence, and this is where you admire
the cleverness of a laterally thinking scientist. Bakker didn't just stop at
his strongly held conviction, but he scoured the evidence of
paleontology, ecology, anatomy, and paleobiogeography for confirming
evidence. Let's go through Bakker's five points.

## ● Bakker's evidence

Bob Bakker listed five main points in support of dinosaurian endothermy:

1) Dinosaurs have complex bone structure with evidence of constant
   remodeling, a bone feature seen in modern mammals, but not in
   reptiles;

2) dinosaurs have an upright posture, as in modern mammals and
   birds;

3) dinosaurs evidently had active lifestyles, or at least the small
   theropods certainly did;

4) predator–prey ratios of dinosaurs show more in common with those of mammals than with those of modern reptiles;

5) dinosaurs are found in polar regions.

## ● The debate

Not surprisingly, there was a furious debate about dinosaur thermoregulation in the 1970s and 1980s – a heated debate, one might even say. Each of Bakker's observations seems pretty watertight, and nobody denied the truth of his five claims. But when paleontologists looked at them more deeply, they found that most of the points were equivocal and did not actually prove warm-bloodedness in dinosaurs. Let's look at the counter arguments.

## ● Bone structure

Close study of the bone structure of dinosaurs shows that, on the whole, they had bone structures like those of modern mammals. When you look at thin sections of dinosaurian bone, they have complex structures full of openings for blood vessels. This suggests active growth. They also have secondary Haversian systems. These are zones that, in cross section, look like fried eggs, and they cut across the primary structure. What is happening is that blood vessels have run through the existing living bone, and they have withdrawn minerals into the blood stream, creating ragged channels through the bone. The minerals are needed during phases of rapid growth. The secondary Haversian systems then fill up with more minerals when the animal is not growing actively.

## ● Mixed bone structures

Close study of dinosaurian bone showed that they can have "warm-blooded" bone in some parts of the body and "cold-blooded" bone in others. Robin Reid, of the Queen's University in Belfast, showed in 1990 that the prosauropod *Euskelosaurus* had typical vascular bone of an endotherm in its limb bones but "reptilian-type" bone with bone rings in its ribs. It can't have had warm-blooded arms and a cold-blooded chest. This example shows that it is not so easy after all to pin down the thermoregulation of a dinosaur just by looking at bone sections.

## ● Haversian systems and large size

The bone of large mammals shows a richly vascular structure and secondary Haversian systems in abundance. Lizard bone does not: it has simple growth rings, laid down, as in a tree, during times of rich food supply. In winter, or when food is scarce, an ectotherm shuts down and stops growing. So surely Bakker was right that dinosaurian bone is that of an endotherm. Well, not really. Small mammals do not show the same bone structures as large ones, and large modern turtles and crocodiles may have secondary Haversian systems. The bone structure is certainly telling us something about the biology of dinosaurs. Bakker was right in that. But it tells us that dinosaurs were big and had spurts of rapid growth. It doesn't tell us that they were endothermic.

## ● Erect and active?

Upright posture does not necessarily mean endothermy. It's true that modern animals which stand fully upright (erect), with their arms and legs tucked under their bodies, are endothermic. But that's based on a sample of two – birds and mammals. Equally, the ectothermic lizards and crocodiles have sprawling postures (see Chapter 5). That's not enough, though. You have to prove a necessary causal link between upright posture and endothermy, and that's not been done.

## ● Active creatures?

Bakker's third point, that at least the small theropods had active lifestyles is suggestive but not unequivocal proof of warm-bloodedness. Insects or small lizards are pretty active, probably more active than most birds and mammals, and they're classic ectotherms. High levels of activity require high metabolic rates (that's a fancy term for the pace of biochemical processes in your body). High metabolic rates usually depend on high body temperatures, but this can be achieved simply by going out on a hot day. No need for endothermy. In any case, fish are pretty active, and most of them have rather low body temperatures.

## ● Predator-prey ratios

Predator–prey ratios for dinosaurs suggest that the predators were endothermic. But what are these predator–prey ratios? They're simply a measure of the relative numbers of carnivores and herbivores in an ecological community (see Chapter 4). Predators are almost always rarer than their prey. If it were the other way round, the whole ecological system would break down. Bakker made a clever observation when he saw that ectothermic predator–prey ratios were about ten times those of endotherms. Why would that be?

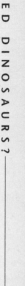

# ● Piles of antelopes

Predator–prey ratios all depend on the amount ectotherms and endotherms have to eat to keep going. Remember that a crocodile weighing exactly the same as you do eats about a tenth of the amount of food. You can picture this in terms of piles of antelopes. Imagine a pile of ten antelopes on one side, and a hundred antelopes on the other. The ten antelopes are what a crocodile might eat in three or four years. The pile of a hundred is what you would eat in the same time (remember, no bread, no vegetables, just antelopes). Turning this around, every hundred antelopes could support ten ectothermic predators (the crocodiles) or one endothermic predator (you). (Pardon me if you're a vegetarian: substitute deer for crocodile and spinach for antelope. It's all the same.) Bakker claimed that dinosaurian predator–prey ratios were nearer one percent than ten percent, a clear endothermic ratio. But there are serious problems in trying to calculate such measures in a precise manner. What are these problems?

# ● Problem 1: taphonomy

The first problem in assessing a predator–prey ratio from an ancient setting is taphonomy. Taphonomy is a word that was invented in 1940 by the great Russian paleontologist Ivan Efremov. He used it to describe all the processes that go on between a plant or animal dying and the discovery of that plant or animal as a fossil. The plant or animal might be eaten, it might decompose, it might be washed along in a river, it might be attacked by weak acids in the soil, it might be crushed by the weight of rock when it is buried and so on. So the big problem in assessing predator–prey ratios is to be sure that the fossil sample is truly equivalent to the living community. Are the beasts all there in the fossil bed in the right proportions and so on?

# ● Problem 2: predation-free herbivores

Another key problem with predator–prey ratios was that some of Bakker's calculations were for settings like the Morrison Formation (see Chapter 4). The main herbivores in the Morrison are the sauropods, such as *Apatosaurus, Barosaurus, Brachiosaurus, Camarasaurus,* and *Diplodocus*. Bakker counted up the skeletons, estimated their weights and worked out the ratios. For the Morrison Formation, it was close to one percent of carnivores (*Allosaurus, Ceratosaurus, Ornitholestes,* and the others) v. herbivores (the giant sauropods, *Stegosaurus,*

*Dryosaurus,* and so on). But wait a minute! Those predators couldn't touch the sauropods (well, only the babies), so, to include them in the calculation will distort it hugely. If the sauropods are excluded, the ratio comes to ten percent predators to prey, an ectothermic figure.

## ● Dinosaurs dancing in the snow?

Surely Bakker was right to point out that polar dinosaurs equaled warm-blooded dinosaurs? Modern reptiles can't live in the polar regions or they'd die. He was certainly right that dinosaurs have been found in regions that lay near the poles in the Jurassic and Cretaceous. Bakker, in 1970, referred to a small fauna of Late Cretaceous dinosaurs from Alaska, then, as now, just inside the Arctic Circle. Greg Paul, a colleague and supporter of Bakker even pictured dinosaurs picking their way through snowdrifts. Surely that must mean endothermy?

## ● Polar dinosaurs from Australia

Since 1970, dinosaurs have been found in South Australia, a region now well out of the Antarctic region. But in the Mid Cretaceous this region lay within the Antarctic Circle (see page 101). The dinosaur fauna from South Australia is made up mainly of small hypsilophodontid ornithopods, one of them, rather appealingly, called *Atlascopcosaurus,* after one of the expedition sponsors. Another of the small dinosaurs, *Leaellynasaura,* showed some details of the brain. A natural rock cast from inside its skull showed that it had extra-large optic lobes. The optic lobes in a reptile brain lie at the front, at the end of longish stalks. Large optic lobes equal good eyesight. And, speculated the discoverers, perhaps this was an adaptation to living in the twilight of the polar winters. Or perhaps not.

## ● No snow

The problem for Greg Paul's vision of dinosaurs in the snowdrifts is that there is no evidence for snow at the poles during the Mesozoic. There are certainly large icecaps over the North and South Poles today, and they keep temperatures north and south at a low level. But without any icecaps at the poles in the Mesozoic, polar temperatures were much warmer, perhaps equivalent to the temperate zones of today. So polar dinosaurs don't necessarily indicate endothermy.

# ● Endless nights

Even if there were no icecaps in the Mesozoic, the polar regions must have been pretty cold in winter. The orientation of the earth and the level of the sun mean that there would have been endless winter nights in the Mesozoic just as today (actualism again). There is some hard evidence for this in the South Australian localities. Some of the lake deposits there show evidence that they froze over in winter. But, as we saw in Chapter 5, dinosaurs could probably have migrated thousands of miles each winter to get away from the cold poles.

# ● A digression on continental drift

I just mentioned that South Australia, today at quite high latitude, was within the Antarctic Circle during the Cretaceous. This is not the place to go into great detail, but it's a fact that the surface of the earth is not stable. The continents and oceans lie over a number of major plates that are all moving ever so slowly – a few inches per year – past each other. Over millions of years, the amounts of movement add up to thousands of miles.

# ● Break up of Pangaea

We can follow the slow story of the dance of the continents through time. At the beginning of the age of the dinosaurs, during the Triassic period (see Chapter 1), all the continents were fused into a single great landmass called Pangaea (meaning "all earth"). During the subsequent Jurassic and Cretaceous periods (see below), Pangaea split up, with the opening of the Atlantic Ocean, North and South. And India, Australia, and Antarctica broke away from South America and Africa and drifted to

**The world in the Early Cretaceous, showing locations of polar dinosaurs, north and south (indicated by dots). The sites in Alaska and in South Australia are best known.**

their present positions. The process continues. Back to the warm-
blooded dinosaur debate.

## ● Sauropod hemodynamics

The warm-blooded vs. cold-blooded dinosaur debate didn't stop at the
discussion of Bakker's five points. Further debating points came up after
1970. One of the first concerned the giant sauropods. With such long
necks, they must have had problems pumping blood up to their heads.
This assertion cannot be questioned: with a neck 16 to 33 feet long,
they had a long way to pump blood to the brain. When the neck is
horizontal, there's no problem. But if the neck is raised, there could be
a big problem.

## ● Pumping blood uphill and keeping it there

Bakker's supporters argued that sauropods must have had massively
powerful hearts to pump the blood at high pressure into the arteries so
it would swoosh all the way up to the head. That's more to supply
blood to the jaw muscles than to the brain, I suppose. They drew
attention to a modern analogue, the giraffe. How do giraffes keep a
healthy blood supply running up that long neck? Why doesn't the
blood just all pool at the bottom? Giraffes have invented a clever trick.
Unlike any other animal, they have valves in their arteries; we have
valves only in our veins. The blood makes its way to a giraffe's head in
stages. Pump past valve one, pump past valve two and so on up to the
top. Whether the sauropods had valves in their neck arteries or not, they
must have been able to counteract gravity somehow.

## ● A four-chambered heart?

To pump blood efficiently and effectively up to their lofty heads, the
sauropods must, so the argument goes, have had a four-chambered
heart, just as birds and mammals do. The four-chambered heart is really
like having two hearts working side by side. One is pumping blood
through the lungs to pick up oxygen; the other is pumping the
oxygenated blood on around the body. When the blood has done a
circuit through the body it comes back deoxygenated. It is siphoned
through the heart back into the lung circuit. So you can think of the
blood system as a figure of eight, crossing over in the double heart.
Amazingly, lizards and turtles have a three-chambered heart.
Oxygenated blood from the lungs and deoxygenated blood from the

body mix together in a single ventricle. Dinosaurs could not have afforded this inefficiency, so they must have had an endotherm's four-chambered heart. But, then, crocodiles have an essentially four-chambered heart. So there's no clear correlation.

## ● The plates of *Stegosaurus*, a radiator?

*Stegosaurus* had a row of tall plates down its back. These might have had a function in defence or in scaring off enemies (see Chapter 8). But they were almost certainly used also in heat exchange. The bony structure of the plates is roughened, showing that it was covered with skin in life. And there are grooves around the base, where blood vessels ran over the bone. If *Stegosaurus* was hot, or if it wanted to look scary, it could pump blood into these vessels and flush its plates red. Heat would be radiated. Likewise, on a cold day *Stegosaurus* could cut down the blood flow to the plates to conserve heat.

## ● Ceratopsian frills as radiators too?

Looking around the various dinosaur groups, the ceratopsians also have a structure that might have been a radiator. Ceratopsians are the dinosaurs like *Protoceratops* and *Triceratops*, that had a broad bony frill at the back of the skull. This bony frill was obviously covered with skin, but probably not much else. Perhaps the ceratopsians were able to increase the blood supply over the frill and cut it off, so they could radiate, or conserve, heat.

## ● The sails of *Spinosaurus* and *Ouranosaurus*

One of the classic dinosaurian images shows two Early Cretaceous dinosaurs from North Africa, both in a desert setting, and both sporting sails on their backs. The dinosaurs are *Ouranosaurus*, an ornithopod related to *Iguanodon*, and *Spinosaurus*, an unusual theropod. Their sails are made from skin stretched over elongated spines on their vertebrae. The idea is that these two dinosaurs lived in such hot conditions that they had to have a radiator system, so predator and prey could sit around gasping in the heat, blasting excess heat from their back sails. But this is almost certainly nonsense. First, just because the bones of *Ouranosaurus* and *Spinosaurus* have been found on the fringes of the Sahara Desert doesn't mean that conditions in the Cretaceous were just the same. In fact, the dinosaur bones are found associated with trees, fish, turtles, and crocodiles. They actually lived in damp tropical

forest. Also, it's not at all clear that the spines did bear a thin sail.
Perhaps they were covered in part by strong back muscles.

## ● Nasal turbinates

Another clever suggestion was made in the 1990s. John Ruben, an
American physiologist, spotted that dinosaurs didn't have nasal
turbinates. So, he argued, dinosaurs were clearly cold-blooded. He
noted that modern birds and mammals have delicate scroll-like sheets
of bone inside the nasal region of the snout. These scrolls are covered
with membranes in life, which operate to warm up the cold air that is
breathed in, and to cool the warm air that is breathed out. It's really like
a car radiator. (This is why your mother taught you to breathe through
your nose in the cold. If you huff and puff through your mouth, the cold
air rushes straight into the lungs, and the hot air goes out uncooled.)

## ● Who needs nasal turbinates?

Modern reptiles do not have nasal turbinates. But it's no problem for
them since their body temperatures are pretty much the same as the
external air temperature. They can breathe in cold air, and it doesn't
cool them down. Nor do they have to conserve hot air, as we do. Ruben
noticed that dinosaurs do not have nasal turbinates, so they were
obviously cold-blooded. Well, maybe. His opponents noted that the
turbinates are immensely delicate, and they might easily be lost from a
fossil. But the challenge is there: if someone can find nasal turbinates in
a dinosaur, then we'll have to reconsider, until then we have to assume
that they lacked these structures.

## ● Air sac systems in sauropods

It might seem hard to imagine at first, but paleontologists have debated
the breathing system of dinosaurs. Did they have a high-powered
system, like birds and mammals, or a slower system, like modern
reptiles? One clue is in the bones of sauropod dinosaurs. Their
vertebrae, the elements of the backbone (see Chapter 1), have great
hollow spaces along the sides. This is partly for saving weight. But, it's
been suggested, the caverns and hollows along the backbone may have
accommodated an air sac system. Birds have air sacs along their
backbones and in their wings. The air sacs are joined by narrow tubes to
the lungs, and they act as overflow systems to increase the oxygen
supply. This might indicate that the sauropods were endotherms. But

it's far from clear. The idea of air sacs in dinosaurs is itself controversial, and even if sauropods did have air sacs, this might be as much to do with their huge size as with anything else. A very different breathing model was proposed in 1998.

## ● A diaphragm

John Ruben argued that he had found a way of determining whether dinosaurs had a diaphragm or not. As everyone learns in school biology lessons, the diaphragm is a broad muscle beneath the lungs. In deep breathing, you pull the diaphragm down and expand the lungs. When you breathe out, you let go, and the diaphragm boings back upwards passively. The idea is that a diaphragm is needed for active high metabolic rate breathing. Modern reptiles don't have a diaphragm, but it doesn't really matter, since they don't need so much oxygen so rapidly. And dinosaurs? Well, the diaphragm isn't preserved, but Ruben spotted another internal organ in a little dinosaur from southern Italy, and that gave him a clue.

## ● *Scipionyx*, **exceptional preservation of soft parts**

In about 1990 an astonishing little dinosaur was found in a limestone quarry in southern Italy. The skeleton was pretty well complete, but it was little bigger than a large seagull. On close examination, the paleontologists spotted an amazing array of internal organs. The whole intestine is there, just behind the ribcage. You can follow it as it runs back from the stomach and curves round, under, over, back and forward, and finally back between the legs. The preservation is so detailed you can even see the bands of circular muscles around the intestine wall that assisted in pushing the digested waste material through. This little beast was named *Scipionyx*, after the famous ancient Roman consul Scipio Africanus, by two young Italian paleontologists, Cristiano dal Sasso, a student in Milan, and Marco Signore, a student in Bristol. But John Ruben wasn't interested in the guts of *Zcipionyx*. He saw something else.

## ● Liver and lights

Ruben spotted what he thought was the liver in the skeleton of *Scipionyx*. Just in front of the intestine, he saw a patch of purplish material. On close inspection, this didn't seem to have any clear form. But it was in the right place to be a trace of the liver. And the purplish color is what you would expect. The liver contains so-called bile pigments (these are what make feces brown). If you eat offal you'll know that livers are purplish. If the purple patch was the liver of *Scipionyx*, then Ruben could determine whether this dinosaur had a diaphragm or not. He argued that the liver was too far forward, virtually within the ribcage, for there to have been a diaphragm. No diaphragm means crocodilian breathing and cold-bloodedness. It's a fairly long line of reasoning, and with plenty of weak points, but who knows?

## ● Isotopes and core temperatures

One of the wildest pieces of lateral thinking in the warm-bloodedness debate came in 1994. In that year, two American geologists proposed a method to measure the body temperature of *Tyrannosaurus*. They applied a well-known geochemical technique, oxygen isotope analysis, a method that allows geologists to assess temperatures by recording the nature of the oxygen locked up in certain minerals. The geologists, Reese Barrick and William Showers, measured the oxygen isotope content of bones from deep within the skeleton (ribs, vertebrae) and of bones from outlying portions (limbs and tail). The measurements suggested that core body temperatures in *Tyrannosaurus* were 40°F (4°C) hotter than peripheral temperatures. Does this imply warm-bloodedness? Perhaps. But it's a wildly daring approach and could be subject to all kinds of errors.

## ● Consensus? Dinosaurs were warm-blooded, but...

It is not clear whether there is any consensus on the warm-bloodedness vs. cold-bloodedness debate. The evidence is so varied. But perhaps it would be true to say that eighty to ninety percent of dinosaurian paleobiologists would now argue that dinosaurs were warm-blooded, but not quite in the way that Bakker had intended. The key seems to be the fact that dinosaurs were so huge that their body temperatures would have been pretty constant day and night.

## ● Alligators large and small

The clue to resolving the dinosaurian metabolism debate came from a piece of classic physiological work published in 1946. Ned Colbert, eminent American vertebrate paleontologist, and his physiologist colleagues Cowles and Bogert decided to look at how the body temperatures of American alligators varied through a range of normal air temperatures. They found that the core body temperatures of small alligators varied pretty much in line with air temperatures. But in larger alligators there was a damping effect. As the air temperature cooled in the evening, the larger alligators cooled only slowly. Colbert and his colleagues found that giant alligators, those weighing about 220 pounds, kept pretty constant core body temperatures day and night, even though air temperatures varied by 68°F (20°C) or more.

## ● Mass homeothermy

A dinosaur that weighed 220 pounds or more would have had a constant core body temperature, whatever the air temperature. Indeed, a large dinosaur would take weeks to cool down, or to warm up, as climates changed during the seasons. It all has to do with heat transfer in and out of large objects. A dinosaur is like a massive heat storage tank: heat it up over a month or two, and it will stay warm for weeks, even during icy cold nights. This seems logical. If you can maintain a constant body temperature – homeothermy – just by being big, why struggle to find enough food to fire the internal endothermic furnaces? Perhaps the big dinosaurs had the best of both worlds: the efficiency and high activity levels of a homeotherm coupled with the low food and water requirements of an ectotherm.

## ● Endothermic small dinosaurs?

The conclusion that large dinosaurs were mass homeotherms does not resolve the question of the smaller dinosaurs. Remember that baby dinosaurs weighed only two to four pounds or two when they hatched, and some theropods, such as *Compsognathus*, weighed only about 20 pounds. No mass homeothermy there! It's certainly likely that these small theropods were endothermic. They would have had to be to keep one step ahead of their larger, mass-homeothermic competitors.

## ● Switching the endothermy on and off

Baby dinosaurs were probably endothermic, at least for a while. At one time, such a suggestion might have seemed impossible as surely you have to be either an ectotherm or an endotherm? But as we've seen, that's not necessarily the case. Dinosaur hatchlings could have started out as endotherms, eating voraciously for a few months, until they became big enough to achieve mass homeothermy. Then they could cut back on the food intake and enjoy life.

CHAPTER SEVEN

# FAMILY LIFE

It may seem highly speculative to consider how dinosaurs ate, moved, and metabolized, but I hope that the discussions in Chapters 4, 5, and 6 show that a great deal can be learned from the astonishingly good fossil record, and from sensible comparisons of living animals and dinosaurs. But surely behavior is out of bounds? Can paleontologists really tell anything about dinosaurs growing up, their parental care and sex? The study of dinosaurian sociobiology must seem to be the wildest of guesswork. But perhaps it is not. Read on.

## ● Did dinosaurs lay eggs?

Dinosaurs laid eggs like modern birds and reptiles; they did not produce live young, like modern mammals. We know that dinosaurs laid eggs because hundreds of fossil dinosaur eggs have been found around the world. Some of the eggs are even in nests, and some of the nests have skeletons of the mother dinosaurs close by. One recent find from Mongolia even has the mother *Oviraptor* actually sitting on her eggs. More of her shortly.

## ● Sausage-shaped eggs

Dinosaur eggs were mostly oval, and some were even cigar-shaped. Modern reptiles lay sausage-shaped eggs or spherical eggs, but dinosaurs do not seem to have laid eggs shaped like baseballs. Some dinosaur eggs have a tapered end, and this seems to have been to stop them rolling out of the nest. They were laid in rough circles of ten or twelve eggs, with the more pointed end inwards. If an egg rolled around a bit, it would always settle back into its place in the nest.

## ● Dinosaur eggs were tiny

Hens lay eggs that are about three inches long, and ostriches lay eggs as big as a soccer ball, maybe ten to twelve inches long. Hummingbirds, on the other hand, lay tiny eggs, less than an inch long. Egg size is clearly related to body size in modern birds. Does this mean that a dinosaur that weighed fifty tons laid an egg as big as a car? No. The biggest dinosaur eggs known to date are about a foot and a half long, and this might be the biggest they could be. Dinosaurs could have produced bigger eggs, no doubt, but the babies would have been locked inside for ever. As an egg becomes larger, its shell has to become thicker. An ostrich egg has a shell about five millimetres thick, and the biggest dinosaur eggs had even thicker shells. But if the shell is a half an inch thick or more, the hatchling would be stuck inside and couldn't crack its way out. Small eggs mean fast growth after birth: more of this in a moment.

## ● Are they like modern birds' eggs?

Dinosaur eggs look like modern birds' eggs, although some of them had strangely pimpled and rippled outer surfaces. They were made of calcium carbonate, a crystalline form of lime or chalk, with the crystals arranged in a radial way – i.e. lined up from the inside out. Modern birds' eggs are just the same. Other kinds of calcite eggs – for example, those of turtles and lizards – have different crystal patterns. Since birds and dinosaurs have pretty much the same crystalline arrangements in their eggshells, it suggests something about the dinosaur–bird relationships (more of this in Chapter 9).

## ● Dinosaur embryos

We know that the dinosaur embryos reached quite a large size inside the egg before they hatched: fossil embryos are scrunched up, and they would have been able to straighten out to twice the length of the egg when they hatched. The fossils don't show it, but dinosaur eggs must have contained a yolk sac. Yolk, the yellow fluid in a bag inside the egg,

**A dinosaur embryo, curled up inside its egg close to hatching. This reconstruction is based on embryos inside eggs from the Cretaceous of North America.**

is rich in protein and fat, and the embryo needs it for growth. Other structures inside a modern bird's or reptile's egg include a sac, to collect waste material, and protective membranes. These were almost certainly all there inside a dinosaur egg.

## ● A private pond

A reptile's or bird's egg can be described as a "private pond," since it's a completely enclosed, safe little aquarium for the embryo to develop in. A whole range of animals lay eggs with shells: reptiles, birds, and (primitively) mammals. Mammals, of course, mainly produce live young from the mother's uterus, but the living duck-billed platypus and echidna, two very primitive mammals, still lay eggs with shells. The shell is semi-permeable: this means that oxygen can pass in and carbon dioxide out, but water cannot pass through, so the embryo does not dry out. Then, inside the egg is a series of bags and membranes. The amnion protects the embryo. The yolk sac, of course, contains nourishing yolk, while waste liquids pass into the allantois. The whole lot is then enclosed in the chorion, which lies just inside the shell, a final protective layer.

## ● Dinosaur nests

I recently asked a class of schoolchildren, "Did dinosaurs make nests in trees like modern birds?" A boy answered, "No, because they couldn't climb trees." Well, yes. But the answer, of course, is supposed to be that (a) a dinosaur would crush any tree it tried to climb, and (b) dinosaurs didn't fly. Nests in trees are not a bad idea: they lift the eggs and embryos out of the reach of many ground-based predators. Ground-nesters today, such as sea turtles, which make their nests in the sand on the beach, lose huge numbers of their eggs and young to scavengers.

## ● Nests like paddling pools

Dinosaur nests were scooped hollows in the ground, something like a child's wading pool. Apparently the mother scratched about to create a shallow hollow, probably digging the nest out by kicking backwards with her hind feet. She then laid her eggs, often in quite a precise arrangement, usually a circle, with the tips of the eggs pointing inwards. Some dinosaurs, certainly some sauropods, seem to have laid their eggs in double parallel lines.

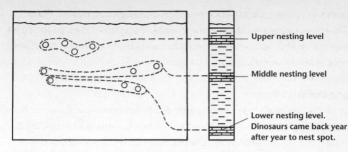

- Upper nesting level
- Middle nesting level
- Lower nesting level. Dinosaurs came back year after year to nest spot.

**Distribution of *Maiasaura* nests on Egg Mountain, Montana, excavated by Jack Horner in the 1970s and 1980s. In the map (left), individual nests (open circles) are spaced equal distances apart. In the vertical section (right), several nesting levels can be identified.**

## ● Where did they build their nests?

Dinosaurs seem to have built their nests in areas of soft dry sand or soil which could be dug easily, and usually close to water. In Mongolia, many nests of a whole variety of dinosaurs have been found in the shore zone of a large lake. In Montana, Jack Horner, a famous North American collector, has found masses of nests at a site called, appropriately, Egg Mountain. Horner claimed that the nest site was on an ancient island, supposedly safe from predators. But it's not clear that it really was an island. In any case, the nest site had to be dry, or the eggs would suffocate (embryos need oxygen), but the nearby water supply may have been needed for the mother.

## ● Egg-laying season

Probably, like many animals today, dinosaurs had mating and nesting seasons. It seems certain that many, if not all, dinosaurs went through mating rituals of one sort or another, where the males postured and growled, trying to win the best mates (see Chapter 8). Rituals of this kind always happen at predictable times of the year. Then virtually all mating would happen in a few days, and so all the eggs would be laid at the same time. Probably spring. Partly because, as they say, "in spring, a young man's thoughts turn to love," but more likely for sound evolutionary reasons. Young dinosaurs that hatched in late spring, when it was warm and when plants were growing at their lushest and animal life was abundant, had the best chance of survival.

### ● Regular nesting spots?

One of Jack Horner's claims about the dinosaur nesting ground on Egg
Mountain, Montana, was that it proved site fidelity. He found nest after
nest as he dug down through yards of sediment. He interpreted this to
mean that the mothers were coming back to this spot year after year
and building new nests over the old ones. Certainly, the site maps and
sections (see page 111) seem to suggest this may be true. It's an
intriguing discovery. Many modern birds, particularly sea birds, may
return to nest at the same spot ten or more times. Behavior like this in
dinosaurs would imply some measure of intelligence.

### ● Spaced out nests and bickering avoidance

Another of Jack Horner's discoveries was that dinosaur nests in popular
nesting grounds seem to be spaced an equal distance apart. The nests
at the Egg Mountain site were built by the duck-billed dinosaur
*Maiasaura*, an animal some 30 feet long. Interestingly, the nests seemed
to be spaced 30 feet apart, too. Horner speculated that this was the
standard "pecking distance" of the dinosaurs. If you look at birds sitting
on the ridge of a roof, or on a telephone wire, they space themselves
equally, just far enough apart that they can't peck each other. Did
dinosaurs do the same? If so, it would mean that, like birds, they were
sociable (living in groups) but bickered a lot.

### ● How many eggs did they lay at a sitting?

Some large sauropods may have laid as few as two to eight eggs at a
time. In southern France, soccer-ball-like eggs of the sauropod
*Hypselosaurus* have been reported in small assemblages arranged in
short double rows. Most dinosaurs probably laid larger numbers of
eggs, perhaps twenty or thirty at a sitting. This certainly appears to be
true for the duck-bill *Maiasaura* and the theropod *Oviraptor*. In these
nests, the eggs are arranged in two or three concentric rings. It isn't
clear whether the mother managed to lay them that way (which would
have been difficult since she didn't have an eye in her rear quarters),or,
more likely, laid them and then rolled them into a neat pattern.

### ● Did they sit on the eggs to keep them warm?

This is another question I ask schoolchildren. They always shout, "No!
They'd crush the eggs." This was true until 1995, but in that year an
amazing specimen was discovered in Mongolia which seemed to show

an adult dinosaur sitting protectively over a nest of eggs. She was apparently killed instantly, possibly by a moving sand dune, and she and her nest of unhatched eggs were perfectly preserved. This is an intriguing, and somewhat ironic story.

## ● The American Museum expedition to Mongolia

In 1923 a major expedition was sent by the American Museum of Natural History to explore Mongolia. The expedition made many remarkable discoveries of dinosaurs, but the most poignant of all was one of the smallest, a number of skeletons of the small ceratopsian *Protoceratops* and some nests full of eggs. When they were put on display in New York they created a sensation, and *Protceratops,* with its nest and hatching babies, is one of the most popular dinosaur icons of the twentieth century. But it turns out that this was all wrong, and that another dinosaur found at the same time had been sadly maligned.

## ● *Oviraptor*: the egg-thief

The American Museum expedition of 1923 brought back skeletons of a lanky animal with long bony fingers and a strange short skull. It was toothless. So what did it eat? The skeletons were found close to specimens of *Protoceratops* and its nests, so the paleontologists decided that the new toothless theropod fed on eggs, and they called it *Oviraptor* ("egg-thief"). *Oviraptor* was an immediate hate figure, the Darth Vader of the dinosaurian era.

## ● *Oviraptor*: the truth

The picture of *Oviraptor* as a beastly egg-thief was well-established. But during a second series of American Museum of Natural History expeditions in the 1990s, it turned out that we had been entirely wrong. Far from being an egg-thief, *Oviraptor* was a caring mother. A complete skeleton of *Oviraptor* was announced in 1995 actually sitting on a complete nest of so-called *Protoceratops* eggs. On dissection, one of these eggs contained an embryo of *Oviraptor,* so the case was proved. The parent *Oviraptor* had her legs tucked in, and her long arms arranged out to the sides as if hugging the nest.

## ● Nest raiders

Many animals would have raided dinosaur nests for a quick and nutritious snack. Even if the toothless theropod *Oviraptor* was not an

egg-thief (and the new find doesn't actually prove that it wasn't –
perhaps *Oviraptor* incubated its own eggs, but it might still have raided
other nests), plenty of other potential predators were about. Today,
lizards, snakes, and mammals commonly raid nests, and some are
specially adapted for a main diet of eggs. There's a strong temptation to
adopt an ovivorous diet: eggs are small packages of protein and fats,
and they don't run away or fight back on the whole. Dinosaur nests
would almost certainly have been raided by mammals and lizards as
well as by small dinosaurs.

## ● How did dinosaurs keep their eggs warm?

Despite the recent find of *Oviraptor* sitting on its nest, it's not clear that
dinosaurs regularly incubated their eggs. Unlike birds, dinosaurs were
heavy, and they would indeed smash their eggs if they lowered their
vast bulk on to the nest. Also, the eggs were often relatively very tiny. A
mother sauropod would be too clumsy to do very much good for her
eggs: it would be like a hen trying to incubate a bunch of snails' eggs.
As far as we can tell, dinosaurs probably covered their nests with sand
once the eggs had been laid, just as turtles do today. Sunlight warms
the sand, and at a shallow depth the temperature would be fairly
constant, but not so hot that the eggs would cook!

## ● Urinating on eggs

Jack Horner found evidence of composting techniques at Egg Mountain
in Montana. Among the sediments over some of the nests, he found
remains of fern fronds and other plant debris. He thought of crocodiles
today. Some of them bury their eggs under great mounds of plant
debris, and then they urinate over the heap. The decaying plant matter
and urine create a nice compost heap that generates heat during
fermentation of the waste matter. This is a highly effective technique,
but probably not such a pleasant introduction to the world for the new
hatchlings as they struggle free through the horrible mess above them.

## ● How long did it take the baby to develop?

It's impossible to be sure how long the embryo dinosaur took to
develop inside its egg. On the whole, smaller animals hatch after a
shorter development time than larger ones. Among modern birds,
incubation periods last from as little as fourteen days for most
songbirds, 20 to 25 days for a chicken, 30 to 40 days for many

seabirds, and 60 to 70 days for the Emperor penguin. So incubation time depends on the body size of the adult (and of the hatchling), and also on the harshness of the climate (hence the very long incubation time among penguins). Perhaps a typical dinosaur embryo was in the egg for 50 to 70 days. But there's probably no way of being able to tell this more precisely.

## ● X-raying eggs

Some of the most exciting dinosaur discoveries in the 1990s were unhatched embryos inside dinosaur eggs. Embryos have been found in the eggs of theropods and ornithopods of several families. Before 1990 some paleontologists had tried X-raying dinosaur eggs but usually they drew a blank. After all, most fossilized eggs were broken, and the hatchling had presumably got out and left the egg behind. In other cases of complete fossilized eggs, the majority were unhatched because they were sterile; they had not been fertilized, and there was never an embryo in them. X-rays showed up shady bones inside some dinosaur eggs, but careful dissection really shows the detail.

## ● Dinosaur embryos

In dissecting a dinosaur egg that might contain an embryo, the paleontologist carefully makes a window into the shell by removing the shell in flakes and then works through the fine sediment inside. This is painstaking work, because the bones of dinosaur embryos are tiny and delicate. They're tiny for obvious reasons. The bones of embryos are especially delicate since they have not fully ossified. Juveniles have a great deal of cartilage in their bones – cartilage is essentially unmineralized collagen (see Chapter 6), the flexible stuff in our ears and nose. But cartilage doesn't fossilize. So the bones of dinosaur embryos are incomplete central pieces. The embryos can sometimes be identified from their bones, but more usually by their associations, if adults are found near the nest.

## ● Dinosaur embryos had teeth

It may seem a little bizarre, but most dinosaur embryos had teeth before they hatched. The embryos may be tiny, but they had a full dental complement along lower and upper jaws. This was essential, of course, since their mothers could not give them milk (only mammals do that). The hatchling had to be ready to eat ferns or worms as soon as it hatched. In some cases the embryos were clearly practicing their chewing moves even before they hatched: there are tiny wear facets on the teeth which show they were grinding on something.

## ● Getting out of the egg: limit to egg size

With a hard crystalline eggshell up to eigth of an inch thick, the dinosaur embryo would have had a hard struggle to get out of the egg. The embryos had a specialized egg tooth, a pointed toothlike structure on the snout, seen in modern reptiles and birds. This is used to crack out of the shell, and it is shed soon after hatching. This is the reason why dinosaur eggs could never be huge. If a 100-foot sauropod dinosaur laid an egg in proportion to the relative size of a hen's egg, it would be 23 feet long. But, at such a size, the eggshell would have to be a foot and a half thick just to prevent the egg from collapsing. The embryo would suffocate inside such a structure, and even if it didn't, it could never crack its way out.

## ● Eating babies

There are social fashions in biology. Victorian travelers sent back tales of the violent and dastardly behavior of the wild animals they encountered. Reptiles came in for a particular bashing. Travelers' tales recounted how crocodile mothers would wait around until their eggs hatched, and then they would lunge in and swallow their own babies. It was only with more careful zoological studies in the 1980s that it was shown that the mother crocodile was actually carrying her babies down to the water for their first swim. Far from acting in a thoroughly immoral manner, she was performing a useful and charitable function. Well, evolution dictates that you help your children. You would go extinct pretty soon if you regularly ate your own babies.

## ● Parental care in dinosaurs

Victorian paleontologists, of course, viewed dinosaurs as even more brutal than the modern reptiles. So it was obvious that they (the dinosaurs, not the Victorians) ate their babies, their brothers, sisters and grannies. But Jack Horner's researches at Egg Mountain, Montana, turned up another remarkable first: evidence for parental care by dinosaurs. Around a *Maiasaura* nest mound, he and his team found fifteen skeletons of small, three-foot-long juveniles. With an adult some 30 feet long, these juveniles must have been only one or two years old.

His suggestion was that they were part of a family group that hung around the nest, helping the new hatchlings.

## ● Helpers at the nest?

If Jack Horner was right, and dinosaur parents, as well as brothers and sisters, hung around the nest, helping the hatchlings, then the dinosaurs were behaving more like birds than like the Victorian caricatures of violent and unpleasant reptiles. Many species of bird have evolved complex forms of family or clan behavior. Sometimes, juveniles born in previous years are active helpers at the nest. The evolutionary explanation for such helping at the nest is that the new hatchlings are brothers and sisters of the one- and two-year-olds. It is in their genetic interest to help such close relatives, and it helps further their line, until they are old enough to leave the group and establish a breeding clan of their own.

## ● Were hatchling dinosaurs helpless?

Among modern birds, hatchlings fall into two classes. These are called altricial and precocial. Altricial hatchlings are born pink and blind, and they require careful attention and feeding for a few weeks before they can fend for themselves. Precocial young, on the other hand, hatch from the egg covered with feathers, with their eyes open, and they are able to trot about and peck for food. Precocial young are seen among quail, ducks, fowl, and most water birds. The presence of parents and older siblings around the *Maiasaura* nests suggests that perhaps these hatchlings were altricial, requiring parental care for a while. But in other cases dinosaur young may have hatched on their own and managed right away without the help of any adults.

## ● Herding and social behavior

Dinosaurs were certainly social animals. So far, we've been assuming that they lived in families, clans, or some other kinds of tightknit groups of related animals. Is this merely sentiment, a kind of hippie backlash against the brutal views of our Victorian naturalist forebears? There is in fact good evidence for dinosaurian social groupings in the form of footprints. Dinosaur tracks give evidence about walking and running speeds and styles (see Chapter 5), and in some cases they can recount hunting encounters (see Chapter 4). But the most common kind of behavior shown by tracks is herding behavior. If you find dozens of tracks of a single species running along side by side, this suggests a monospecific herd moving past. In many cases, individual tracks are of different sizes, indicating that the herds contained adult males and females as well as juveniles. Can we read more from the tracks?

## Guarding the young

One intriguing track from Cretaceous rocks at Glen Rose, Texas, shows size sorting of tracks. The smaller footprints appear to lie between larger prints. Here perhaps was a small herd of sauropods trotting along, with beefy adults on the outside, protecting the juveniles in the center. The evidence is not unequivocal, but it's highly suggestive. This kind of protective social behavior among dinosaurs is now widely accepted. A common image shows a herd of horned ceratopsian dinosaurs, perhaps *Triceratops*, arranged in a circle and facing outwards, the juveniles safe at the center. This type of defensive circle, modeled on the behavior of musk oxen today, is plausible, but it is guesswork. Have we perhaps gone too far in reinterpreting dinosaurs as lovey-dovey, snuggly, friendly creatures?

## How long did it take a dinosaur to reach adult size?

Baby dinosaurs hatched very small, as we've seen. Did they then grow at reptilian sorts of rates, or much faster? At the relatively sluggish rates of growth seen in modern crocodiles and snakes, for example, a medium-sized dinosaur might take fifty to sixty years to reach adult size, a sauropod as much as two hundred years. This is not impossible, but it might seem a very long time. Mammals can grow much faster. An elephant reaches adult size in twenty years, so, at these kinds of rates, a sauropod might have done it in thirty or forty years.

## Counting the bone rings

At one time paleontologists hoped that they could age a dinosaur by counting the bone rings. Well, this might work pretty well for trees, but it doesn't seem to work for animals so well. The science of skeletochronology, as it's called, has for a long time tried to establish a clear link between rings in modern reptile bones and their age. But it doesn't really work. Two lizards of exactly the same age may be different in size, and they may have different numbers of growth rings in their bones. All it takes is a sudden glut of food and the lizard grows. So it might lay down two or three growth rings in a year. It was probably the same for dinosaurs. Counting bone rings in sauropod bone has suggested that they took thirty years to reach a weight of six tonnes. This would imply as long as 300 years to reach the adult weight of 55 tons. Dinosaurs could probably lay down two or three growth rings in a year, so the calculation goes haywire.

## ● Grandchildren = evolutionary success

Is there any reason why dinosaurs shouldn't take 200 or 300 years to reach adult size? The worry about such a long time for growing up is that it is enormously risky in evolutionary terms. The key to success is for your children to reach sexual maturity and to produce their own offspring. In other words, grandchildren mean that your line will survive. Until a dinosaur had itself reached sexual maturity, mated, and produced its own young, it had not contributed to the evolutionary continuation of its line. The longer it takes to reach sexual maturity, the more risk there is of death or injury.

## ● Indeterminate growth in modern reptiles

We are used to the idea of a kind of fixed adult size, since this is seen in birds and mammals. When that adult size is reached, the animal simply stops growing. But many modern reptiles show what is called indeterminate growth: there is no defined adult size, and they can keep on growing, admittedly at a slower rate, beyond the size at sexual maturity. That is why one hears, from time to time, tales of amazingly huge anacondas or crocodiles. These monsters are rare, the adults that were lucky enough not to be killed and which show the full growth potential of the species.

## ● Dinosaur growth patterns

The jury is still out concerning the situation in dinosaurs: did they show determinate or indeterminate growth? Occasionally paleontologists find isolated overlarge individuals, which might suggest "reptilian" indeterminate growth. But mostly there seems to be a "typical" adult size for a species of dinosaur. So maybe they mostly showed determinate growth, as in birds and mammals.

## ● Bone fusion

Embryos have no bones, merely cartilage, as we've seen. So it is possible to age a vertebrate by the rate at which the cartilage is replaced by bone. On hatching, a dinosaur already had bone in the shafts of its limb bones and in other critical areas: for example, around the spinal cord in the vertebral column, and in the main bones of the skull. But the ends of most bones remained cartilaginous for some time. This is true in human babies also. It's an obvious way to allow the juvenile to grow larger: bone is laid down in the middle parts of limb bone shafts, but the bone ends only become fully ossified (i.e. turned to bone) at full adult size.

## ● Baby dinosaurs had big heads and eyes

Human babies have big heads and big eyes, and the same is true of puppies, kittens and indeed virtually all babies, including dinosaurs. Why? The most common reason that is given in books about bringing up your children is that it makes babies appealing to their parents. When they look at you with their big goggly eyes, you just have to pick them up and cuddle them (and preferably feed them). It might be too sentimental to assume that this is the reason for big eyes and big heads in other animals. The key reason is that some parts of the body are more complex than others, and they develop prematurely. Brains and eyes have to be as near to adult size as early as possible so that the baby can learn and function properly.

## ● Baby dinosaurs had big knees

Many baby animals have oversized limb joints – elbows and knees like soccer balls. This is true also of human babies. Young deer, antelope, horses and cattle have almost full-sized adult legs all ready for action the minute they are born. That seems obvious, since they often have to be ready to keep up with the herd right away. But why would human and dinosaur babies have big knees? Some dinosaurs perhaps had to be ready to trot as soon as they were born. Others, though, stayed around the nest. Human babies can't walk when they are born, so why would they have big knees? Knees are big since they are complex joints that cannot keep adjusting their shape as the bones develop and increase in length. The joint seems to be put in place early, and then requires less reshaping during growth, thus avoiding the risk of malformation.

## ● How can you tell a juvenile dinosaur from an adult?

Juvenile dinosaurs are obviously smaller than the adults. The particular proportional differences are good guides as well: relatively large eyes, heads, knees, and so on. Finally, as we've learned, juveniles have more cartilage in their skeletons than adults, and the joints only became fused properly at full adult size.

## ● Dinosaur teeth and growth

In humans and horses, you can age an individual pretty accurately by looking at the teeth. A dentist or a veterinary surgeon is skilled at the art. This is because mammals have two sets of teeth, milk teeth and adult teeth. They are replaced in a fairly predictable way. That produces an aging system up to about twenty years for humans, when the last wisdom teeth erupt through the jaws. Then, the state of wear of the adult teeth can give a hint of approximate age after twenty. This method does not work for dinosaurs, of course, since they had many sets of teeth during their lives, perhaps a new set every year or two. Good news for dinosaurs; bad news for dentists.

## ● Individual variation

In the early years of dinosaur research, every new skeleton that was dug up was given a new name (see Chapters 1 and 3). But then paleontologists began to find whole assemblages of skeletons, maybe ten or twenty, sometimes as many as a thousand, of apparently a single species. They then had to recognize that dinosaurs showed variation just as individual humans do. Some might have relatively longer arms, higher skulls, wider tails and so on. This is called individual variation.

## ● Telling male and female dinosaurs apart

In looking at many individuals of the same species of dinosaur, paleontologists have often noticed that they seem to sort into two morphs ("forms"). The skeletons are identical in every regard, except that there is a so-called "robust" and a "gracile" form. The robust morph has, say, slightly thicker bones than the gracile form, maybe by just ten percent or so. The robust morph is sometimes slightly larger than the gracile morph, but not to any great extent. It's assumed that we are not dealing with juveniles and adults here.

## ● Sex: size *does* matter

If there are robust and gracile forms in a population of dinosaurs, then surely the robust one must be the male? Rough, tough and thick-boned? In fact, the males are just as likely to have been gracile. Males are larger than females among mammals and other groups because they perhaps have to fight for mates, hunt for food, fight off enemies, and so on. But it isn't always the case that males are the larger sex.

## ● Females bigger than males

Among egg-laying reptiles and other egg-laying animals, it's often the female that is bigger because she has to bear the eggs. Female frogs are twice the size of their mates. Some female spiders and insects are behemoths beside their weedy little spouses. The female angler fish is large, and the male is a tiny little creature, little more than a swimming testicle, that attaches himself parasitically to the side of the female. So the robust dinosaurs may have been the females, extra-large so they can withstand the rigours of egg-laying and caring for the young. Who knows for sure?

## ● Duckbilled dinosaur crests

Telling the sexes apart is dramatically easy among the hadrosaurs, the duck-billed ornithopods. The hadrosaurs are known for their amazing range of headgear. Some had pointed crests, ridges like dinner plates, tubular structures, and all kinds of extra structures. Recall, too, that many hadrosaurs did not have such headgear. But among those that did, the shape of the crest can vary. In *Parasaurolophus*, for example, some adults had a very long head spine, while others had one about half the length. It's assumed that one is the male, the other the female. Juvenile hadrosaurs had short little crests, since they were born with nothing much. The crest sprouted at an accelerated rate as they approached sexual maturity. What was the crest for?

## ● Signalling hadrosaurs

The most popular explanation of the hadrosaur head crest is that it was a signaling device. Certainly, each species would stand out, with the great array of crests. The hadrosaurs seemed to have fed in mixed-species herds in

**Skull of the hadrosaur *Lambeosaurus*, showing its head crest. Nasal passages ran up inside the head crest, through to the back of the throat.**

the Late Cretaceous. Several species are often found, preserved together, and distinguishable only by their head crests: their skeletons look the same. So, like us, probably the hadrosaurs were visual animals, spotting members of their own species by the shape of the crest, probably also highlighted by specific coloring patterns. But there's more.

## ● Honking and snorting

The hadrosaur crest is not a horn or spine in the normal sense. It's an outgrowth of the nasal and frontal bones of the skull. The air passages from the nostrils to the throat run up and around inside the crest (see page 122). When a hadrosaur snorted, the noise it made would be modulated by the pattern of winding of the nasal tubes encased in the bone. Each crest shape had a different winding pattern of the nasal tubes. So, suggested American paleontologist Jim Hopson in 1975, a herd of hadrosaurs was like a brass band. Each crest is a different instrument, and the resulting honks, whistles, and hoots would have been species-specific signals. Males, females, and juveniles had different crest shapes, so the honking and hooting could be used between members of families or in sexual skirmishes.

## ● The horns of the ceratopsians

Various dinosaur groups may have used their crests, horns, and spines as sexual signaling devices. The head crests of hadrosaurs are the most obvious example. But perhaps the ceratopsians, with their varied head shields and face horns, might have used them in the same way. Certainly, these structures were probably used in fighting and defence (see Chapter 8), but structures can have more than one function. Covered with a gaily patterned skin, the ceratopsian neck shield would have made a good signaling device. The arrangement of horns on the head certainly indicates the species, and perhaps males waved their heads around as a threat, trying to look as spiky and horny as possible.

## ● Did male dinosaurs fight for their mates?

Almost certainly, before the mating season began, male dinosaurs would have postured and strutted to impress the females. They might even have engaged in head tussling and minor fighting moves. But, like modern animals that do these kinds of things, it was mainly bluff. There's no advantage in fighting to the death if you can avoid it.

CHAPTER EIGHT

# FIGHTING AND DEFENSE

Children love bloodthirsty scenes, and every kid's dinosaur book seems to be full of biting, tearing, and gouging. Is this just an overdramatization or was life in the Mesozoic really like that? The phrase "nature red in tooth and claw" was first used by Alfred Lord Tennyson, the great English Victorian poet, to describe his vision of life in the wild as envisaged by Darwin. Fighting and defense are indeed a critical part of every animal's life, and dinosaurs no more or less than any others. But perhaps there's a bit of selectivity in the books. Certainly, it seems likely that all dinosaurs might have engaged in a bit of pre-mating fighting, but this was just between males, and just for a day or two each year.

## ● The life of a soldier

Dinosaurs almost certainly did not spend their whole lives fighting, like some latter day professional wrestlers. Where the children's books are wrong is in showing only fighting and defense: most of a dinosaur's life was probably very dull, as it went about the business of finding food, resting, twitching annoying flies off its flanks, drinking water from puddles and so on. But they did have to fight and defend or they would have died.

## ● The fighting *Iguanodon*

From the start, artists liked to depict dynamic scenes of dinosaurs in action. A famous picture from a popular Victorian book about the history of life, Louis Figuier's *The world before the Deluge*, shows two dinosaurs tussling. They are rather rhinoceros-like, indeed based on Sir Richard Owen's Crystal Palace models of *Iguanodon* and *Megalosaurus*

**Head-tussling *Triceratops*. Like modern deer, they may have struggled to establish who was boss, but they probably didn't fight to the death.**

(see Chapter 2). They have their teeth sunk into the necks of each other, in a weird kind of unending circle. It's hard to see what their next moves might be. The picture is also rather odd in that *Iguanodon* is, of course, a herbivore, so why it's sinking its blunt little teeth into *Megalosaurus'*s neck is a bit of a mystery.

## ● Why fight?

Fighting is for defense, feeding or to secure a mate. Animals do not fight for fun – although young animals may engage in some rough and tumble while growing up – we leave that for humans alone. Plant-eaters have to be able to defend themselves from predators. Flesh-eating animals have to be able to overwhelm their prey, at least often enough to ensure that they do not starve. And many male animals engage in mock fights for mates. Dinosaurs did all of these things.

## ● Did male dinosaurs fight for their mates?

Almost certainly, before the mating season began, male dinosaurs would have postured and strutted to impress the females. They might even have engaged in head tussling and minor fighting moves. But, like modern animals that do these kinds of things, it was mainly bluff. There's no advantage in fighting to the death if you can scare off your rival by a bit of roaring and bluster. That's why many birds and mammals that do this kind of thing have impressive head decorations – antlers, colored crests, inflatable throat pouches – which they flourish at their rivals in an attempt to intimidate them.

## ● Head wrestling and butting

Several groups of dinosaurs had threatening headgear of one sort or another. Think of the crests of the duck-bills, the face horns, and neck frills of the ceratopsians, the spines, and plates along the backs of the stegosaurs, and the spiny knobbly covering of the ankylosaurs. In about 1960 the Russian zoologist L. S. Davitashvili showed how modern mammals engage in posturing and mock fights for mates, and paleontologists soon applied his ideas to dinosaurs. The assumption is that, in all cases, the stronger male (or the better show-off) gets to mate with the female, or often females, and the loser does not. So there's a strong evolutionary imperative at work here, known as sexual selection.

## ● Ceratopsian head wrestling

Ceratopsians clearly used their neck frills and face horns in defense, but these structures also would have been ideal for pre-mating mock fights. Today, for example, male red deer lock antlers and push and shove. Usually, the weaker one is exhausted after some minutes of this exertion and runs off. He may receive some superficial wounds, but it's almost unknown for the loser to be killed. A skull of *Triceratops* in the Science Museum of Minnesota, in St Paul, has a deep dent in the side, just below the eye socket. The dent is the right size and shape to have been caused by the nose horn of another *Triceratops*, so here perhaps was a fight that went a bit too far. But the injured adult may well not have been killed by the wound.

## ● Head butting dinosaurs

Today, wild sheep and goats engage in head butting as a trial of strength between rival males. And there were head-butting dinosaurs too. The pachycephalosaurs are a strange little group of mainly Late Cretaceous dinosaurs, known from North America and central Asia. They look pretty much like a standard ornithopod such as *Hypsilophodon* or a lightweight *Iguanodon*, but their skull roofs are immensely thick, sometimes as much as nine inches thick in a skull two feet long. Obviously, these dinosaurs head-butted just like modern mountain sheep right?. Well, maybe not: there is an ongoing debate, and the problem is that their necks do not seem strong enough to absorb the huge forces of such impacts. But the opponents of head butting still have yet to explain the thick skulls. I would still go for head butting in pachycephalosaurs.

## ● Weapons of the theropods

All flesh-eating dinosaurs are classed in the group Theropoda and, so far as we know, there were no flesh-eaters among the other dinosaur groups. However, recently it's been realized that some of the theropods might have graduated to an herbivorous diet, but only a very few. More of those in a moment. The point is that, throughout the 165 million years of the age of the dinosaurs, theropods were the predators, small, medium, and large, and they had to rely on their wits and their weaponry to survive.

## ● Theropod teeth

The key to the success of the theropods was their teeth (see Chapter 4). Almost all theropods inherited the standard flesh-tearing teeth of their Triassic ancestors, and they stuck with them. It was an amazingly persistent design. *Tyrannosaurus*, perhaps the biggest land predator of all time (but there are some contenders, see below), had teeth that were almost identical in shape to those of its ancestors from 165 million years before. We've seen in Chapter 4 that the theropod killing teeth were flattish-sided, with a sharp edge fore and aft lined with sharp serrations like a steak knife. The teeth also curved backwards so that a struggling prey animal could not escape outwards; it could only struggle itself further and further into the throat of its assailant.

## ● Killing by bacteria and bad breath

Careful consideration of theropod hunting techniques has shown that perhaps they killed more by bad breath than by tearing or stabbing their victims to death. Modern large predators do this. Their teeth are covered with shards of flesh from previous meals, and this is a breeding ground for bacteria. Wild cats and dogs can have pretty bad breath. But when a predator bites another animal, the bacteria are transferred and infect the wound. If bleeding does not kill it, the infection may. Theropod teeth, with their fine-edge serrations, were an ideal design to foster the proliferation of bacteria. Not so dramatic, perhaps, as tearing chunks of flesh from your prey, but death by wound infection is just as effective and less risky.

## ● Toothless jaws

Of course, not all theropods had teeth. As we saw in Chapter 4, some, such as the troodontids, had only tiny teeth, and others, such as the ornithomimosaurs and oviraptorosaurs, had no teeth at all. Their diet has been debated. Because they had small teeth, or no teeth, many paleontologists believe they had special diets, such as small items of prey, eggs, or even plants. But no teeth, of course, does not mean they could not have been effective predators. Today, birds of prey cope perfectly well with a full range of prey items, and they find their sharp-edged beaks are just as effective as teeth.

## ● Hands and claws

The other main weapons of theropods were their hands and feet, especially the hands. All theropods were bipeds, and they remained bipeds, even the five- to ten- ton behemoths like *Tyrannosaurus*. This allowed them to use their hands and arms for purposes other than locomotion, and most theropods had strong sinewy arms equipped with long fingers and sharp claws. The smaller theropods at least almost certainly used their hands in subduing prey, and all theropods would have used their hands in dismembering a carcass.

## ● The fish-hook of *Baryonyx*

The Early Cretaceous theropod *Baryonyx* from near London, England, is famous for its huge heavy claw (the name *Baryonyx* means "heavy claw"). It is reconstructed with the claw on one of the fingers of its hand, but it's not absolutely certain that that is right since the claw was found lying beside the rest of the skeleton. In any case, *Baryonyx*, like the African spinosaurids, had a very strange long-snouted skull, with rather feeble jaws and tiny teeth. Inside its ribcage were found some fish scales. So, speculated the discoverers, perhaps this was a fish-eater. Perhaps *Baryonyx* used its large claw to flick fish from the river, just as bears do today.

## ● *Alxasaurus* scissor-hands

One theropod group definitely gave up flesh-eating in favor of vegetarianism, the therizinosaurids. These bizarre animals were like a cross between a ground sloth, an oversized gorilla, and Edward Scissorhands (see page 129). Their vast dangling arms were equipped with incredible long claws. For decades, only bits and pieces of the skeleton of the therizinosaurids were known, and no one had a clue what they were. Now, even with essentially complete skeletons of *Alxasaurus* from northern China, they're still a mystery. The claws look much like hunting weapons, and yet they must have been used simply for gathering branches to eat.

◀ *Coelophysis*, a small predator from the Late Triassic of North America, lived in large groups. These dinosaurs probably squabbled and bickered all the time.

▲ Dinosaurs on the move. A herd of ornithopods, *Muttaburrasaurus* from the Early Cretaceous of Australia, head through the dusty territory for richer feeding grounds.

▶ Many dinosaurs seem to have been good parents. A mother *Leaellynasaura* from the Early Cretaceous of Australia about to regurgitate some pre-chewed ferns for her babies.

▲ The most extreme
dinosaurian killing
machine; *Utahraptor*,
armed with a slashing
claw on its foot.

◄ Attack and defense.
The predator *Allosaurus*
is confused by a display
of flushed defense
plates by *Stegosaurus*.

◀ **Guarding the family.** The juvenile cerutopsian *Torosaurus*, from the Late Cretaceous of North America, sticks close to his mother for protection against predators.

▲ **Like a tank on legs,** *Ankylosaurus* from the Late Cretaceous of North America, had two lines of defense. His armor-plated body and bony tail club kept predators at bay.

▶ **The most fearsome dinosaur of all,** *Tyrannosaurus rex* was also one of the last dinosaurs on Earth. *Tyrannosaurus rex* was the biggest terrestrial carnivore of all time, but was he an active hunter or a scavenger?

A large flesh-eating theropod, *Sinraptor* (left) menaces a pair of herbivorous therizinosaurs, *Alxasaurus*. These weird theropods had heavy, slow-moving bodies, and long scythe-like claws, whose function is still a mystery.

## ● The rapid raptors

A small dinosaur found in Alberta, Canada in about 1920 was the first of a deadly new group, but this was only fully realized fifty years later. *Dromaeosaurus* was named in 1922 by the American paleontologists William Matthew and Barnum Brown, based on an incomplete skull and odd bits and pieces of the skeleton. Only after the discovery of much more complete skeletons of another dromaeosaurid, *Deinonychus*, in 1964 did John Ostrom show the world what the raptors were really like.

## ● *Deinonychus* and a new way of hunting

*Deinonychus* was only six feet long, but it was armed and dangerous (see page 35). The arms were long, and each of the three strong bony fingers was equipped with a sickle claw. The same applied to the toes, but the second toe had a double-sized claw. Ostrom argued that this was a hunting, slashing claw, and it worked something like a flick knife. More of that in a moment.

## ● Cats and dogs

Until Ostrom's description of *Deinonychus* in 1969, most paleontologists had pictured the theropods as essentially lone hunters, like grumpy, solitary crocodiles or hunting cats. But there is another hunting style, shown today by several groups of mammals. For example, wolves and African hunting dogs operate as packs. They select a prey animal often much larger than themselves, and harry and worry it until it falls, exhausted. So a relatively small predator can hunt larger prey with a little teamwork. Of course, another mammal that does this is *Homo sapiens*. Ostrom's rather startling insight was that *Deinonychus*, and the other raptors, probably hunted in this way. And there is fossil evidence in support of his idea, in the form of a kill site (see Chapter 4).

## ● The sickle claw

The sickle claw on the second toe in *Deinonychus* was four inches long, and that in an animal the size of a ten-year-old child. In detail, the second toe is immensely flexible. Unlike our toes, each of the bony elements of the toe could be bent upwards, especially the joint between the claw and the two joints immediately behind. So when it was running, *Deinonychus* folded its huge sickle claw back, well clear of the ground. But it had powerful muscles underneath the toe. It could leap at a prey animal and swing the claw down through 180 degrees in an instant, creating a deep slash up to a yard long.

## ● Double-sized *Velociraptor*

*Deinonychus* and *Dromaeosaurus* from North America are closely related to *Velociraptor* from Mongolia, all three being dromaeosaurids, or raptors. *Velociraptor* is famous since it starred in Steven Spielberg's *Jurassic Park* in the famous kitchen scene. Spielberg decided, for filmic reasons, to make *Velociraptor* twice its actual size – it was actually a mere five feet long – so that it would seem more scary. Almost at the same time as the film was shown (1993), a double-sized raptor was unearthed in North America, now called *Utahraptor*. So, clearly you should believe everything you see in the movies!

## ● The fiendish troodontids

The troodontids were close relatives of the dromaeosaurids, although more slender and less well known. In fact, although these are some of the most fascinating dinosaurs, for a rather unexpected reason, as we'll see, there are only about twenty rather fragmentary specimens known, from the Late Cretaceous of North America and Mongolia. The troodontids share with dromaeosaurids a sicklelike hunting claw on the foot, though it is not as large as the dromaeosaurids'. But they may have hunted in packs in the same way. Troodontids are distinguished among dinosaurs by their intelligence.

## ● Humanoid dinosaurs?

There is no doubt about the intelligence of the troodontids. Unlike all other dinosaurs, they have a bulbous braincase, more bird than reptile. In part, the brain was enlarged in the sensory regions: so troodontids had large eyes and a good sense of smell, as well as a good sense of hearing. But, it's been speculated, improved senses and pack hunting mean more actual intelligence. In 1982 the Canadian paleontologist Dale Russell speculated about what might have happened had the dinosaurs not died out 65 million years ago. Perhaps the troodontids would have inherited the earth, becoming ever more humanoid. He and sculptor R. Séguin created a model of a dinosaurian humanoid, based on the projected evolution of troodontids through another 65 million years. It has a bulbous cranium, large goggly eyes, and it walks upright without a balancing tail. Scary concept.

## ● Dinosaurian intelligence

On the whole, dinosaurs are not noted for their intelligence. When the gray matter was being handed out, as they say, dinosaurs were pretty well last in line. But how can we be sure? Well, intelligence depends roughly on relative brain size. First you assess the absolute brain size, and then you divide it by the body weight to find the *relative* brain size. You can measure the volume of an animal's brain from the size of the braincase cavity. Paleontologists do this quite simply. They fill the braincase with dried lentils, and then pour the lentils into a measuring cylinder. The brain volume is read in cubic centimeters (cc). Divide this figure by the body weight, and you can then compare relative brain sizes among different animals.

## ● Reptilian intelligence levels

Dinosaurs had relatively tiny brains. In proportion to body size, their brains were a tenth, or less, of the relative size of the brains of mammals, for example. But, for reptiles, their brains were not excessively small. In other words, relative to their body weights, the brains of *Tyrannosaurus* and *Brachiosaurus* were the right size for a reptile. This clearly means they could only do the things that modern reptiles can do, and it is important not to try to make dinosaurs perform enormously complex behavior.

## ● A brain like a walnut

Which was the most stupid dinosaur? The answer is *Stegosaurus*. With a body weight of six to seven tons, *Stegosaurus* had a brain the size of a walnut. Its brain was the same size as a kitten's, but a kitten weighs only two or four pounds, so the difference in the relative brainpower is amazing. We can say, with some confidence, that the brainpower of *Stegosaurus* was at best one three-thousandth that of a kitten (2:6000). Pretty hopeless really!

## ● Dinosaur eyesight

Dinosaurs had eyes, so they could see. But how well? Could they see color? How precise was their vision? Could they see in three dimensions, or only two? We probably can't resolve many of these aspects for dinosaurs, except the 2-D *v.* 3-D question. The secret of three-dimensional vision is to have eyes on the front of the face, so that the fields of view overlap. It's by seeing a single object from a slightly different angle through each eye that it is given its 3-D qualities.

## ● Three-dimensional vision

The secret of three-dimensional vision is to have overlapping fields of vision. Modern horses and dogs cannot do this since they have a long high snout, with an eye on each side. To see what's in front of them clearly, they have to nod their heads from side to side so that each eye gets an image. Using this principle, it turns out that many dinosaurs *did* have overlapping fields of vision, so they could have seen in 3-D. This is true especially of the hunters, which had to be able to judge distance precisely. Imagine a *Deinonychus* with 2-D vision: it would leap at its prey animal and either fall short or go sailing right over its back. Definitely 3-D vision in the theropods.

## ● Dinosaur hearing

On the whole, dinosaurs could probably have heard reasonably well, but maybe not much better than we can. They had well enough developed hearing regions in their brains, but not excessively. And their ears were small, as in living reptiles. We have no evidence that they had the huge sound collecting devices of mammals, the external ears. Perhaps they did – imagine a *Diplodocus* or a *Tyrannosaurus* with big flapping elephant ears – but living reptiles and birds don't, so they probably didn't either.

## The carnosaurs

The large theropods, such as *Allosaurus* and *Tyrannosaurus,* used to be grouped together simply because they were large. The large theropods were called generally the carnosaurs ("flesh reptiles"). These animals relied on their large size to allow them to hunt large prey animals, and they did not seem to have any of the fiendish adaptations of acute eyesight, slashing claws, or high speed seen in the smaller theropods. New studies of dinosaurian relationships (see Chapter 9) now suggest that large theropods evolved in several separate lines, so the term "carnosaur" has only a broad ecological merit, but in fact has no evolutionary value.

## Scary crests and horns

It wasn't just the plant-eating dinosaurs that had crests and horns for signaling and fighting. Some theropods did, too. *Ceratosaurus* from the Late Jurassic Morrison Formation of North America had a pair of bumps in front of its eyes. *Dilophosaurus* from the Early Jurassic of Arizona had a pair of raised ridges along the midline of its skull. Some of the Late Cretaceous predators from South America, such as *Carnotaurus*, also had crests and knobs on the skull. It would hardly seem necessary to use these to scare their prey: the sight of the sharp teeth and vicious claws would be quite enough. So these structures might have been used when males postured and blustered, trying to make their rivals back down. Some of the knobs and crests contained air spaces, and they might also have acted as resonating chambers to help create a scary roar.

## Hunting *Apatosaurus*

In the Morrison Formation, bones of sauropods such as *Apatosaurus* often contain bite marks that match the teeth of *Allosaurus* and *Ceratosaurus*, the two large carnivores of the day (see Chapter 4). Clearly the sauropods were hunted and attacked. But such finds are rare, and this may have been a rare occurrence. Or were the carnivores merely scavenging a carcass? This issue is important for two reasons. The first is for understanding the hunting behavior of large theropods: were they active hunters or sluggish scavengers? Secondly, the issue of whether sauropods formed part of the diet of Jurassic theropods or not is critical for calculating predator–prey ratios (see Chapter 6).

## ● *Tyrannosaurus*, the biggest flesh-eater

When *Tyrannosaurus* was named in 1905 by Henry Fairfield Osborn, he
was an eminent man. President of the American Museum of Natural
History, Professor Osborn was a proud, autocratic man. Here was
another triumph for him, and for his institution: the biggest flesh-eating
dinosaur of all time, 40 feet long and weighing seven to ten tons.

## ● Contenders from Africa and South America

*Tyrannosaurus* held the record as biggest flesh-eater for most of the
twentieth century. But challengers from other parts of the world bared
their ugly fangs in the 1990s. First came *Giganotosaurus* from the Mid
Cretaceous of Argentina, and then *Carcharodontosaurus* from North
Africa. Their discoverers claimed they were bigger than *Tyrannosaurus*.
And while they were arguably a bit longer, in the end it had to be
conceded that *Tyrannosaurus* was heavier, and therefore more scary.
Its pre-eminence was confirmed in 1997 with the sale of a new skeleton
for $8 million and the offer of a second new specimen in 2000 for
$15 million. Checkbooks out!

## ● Active hunter or scavenger?

Dinosaur experts are divided into *Tyrannosaurus* fans and *Tyrannosaurus*
apologists. The fans see this vast biped striding about the Late
Cretaceous landscapes of North America at high speed, running down
prey, jumping on the backs of hadrosaurs, and tearing them to shreds.
The apologists point out that, although it was big, *Tyrannosaurus* had
weedy little arms and only two fingers on its hands: it couldn't even
reach its mouth. They also note that it was so huge, and its safety
factors so marginal, that it couldn't go faster than a leisurely amble
(see Chapter 5). So *Tyrannosaurus* couldn't hunt for nuts: he just ate
whatever he stumbled over. And that's another thing: recent
calculations suggest that if *Tyrannosaurus* did trip up he'd be so winded
by his vast weight that he would die. The debate continues: was
*Tyrannosaurus* a terrifying hunting machine or a lumbering wreck?

## ● Injury and disease

Dinosaur skeletons usually seem to come from healthy animals. But
every so often, signs of injury and disease can be identified. These are
the subject of a new science called paleopathology. Paleopathologists
spend most of their time looking at ancient human skeletons working

out the kinds of diseases our ancestors suffered from. But dinosaur skeletons show broken bones, healed fractures and wounds caused by carnivore teeth and by fighting (see the *Triceratops* example above). Dinosaurs, like humans, suffered from arthritis (where the bone joints rub painfully and become overgrown), malformations of bones caused by traumas of one sort or another, inflammations, tumors, fusion of elements of the backbone, and even gout.

## ● Fight or flight

With all the weapons and guiles of the theropods, it might seem astonishing that any of the plant-eaters survived. But that's nature. There always *have* to be more prey than predators. If a predator became too efficient and greedy, and destroyed all its prey populations, it wouldn't last long itself. The plant-eating dinosaurs had two options when they were attacked: fight or run.

## ● Speedy running to escape

Running away was probably the most common defense against the theropods. Indeed, for many plant-eating dinosaurs, it was their only option. The smaller ornithopods, such as *Hypsilophodon* and *Leaellynasaura*, were small and had long slender legs. They could probably have outrun any predator unless they were caught unawares. Even some of the medium-sized plant-eaters, such as the larger ornithopods (*Iguanodon*, hadrosaurs), the stegosaurs, and the ceratopsians, would take to flight, and they could trot, if not gallop, at a fair speed for a short time.

## ● Prosauropod thumb claws

Prosauropods such as *Plateosaurus* from the Late Triassic of Germany were generally quite defenseless. But there weren't many large predators around then. They did have large thumb claws, which might have been used for raking in branches to eat but also could have

**The hand of *Diplodocus*, showing the five toes, the middle one (the thumb) armed with a long claw. The others have small hooves. Why would a massive plant eater have a sharp thumb claw? Perhaps it could have been used in fighting.**

deterred some predators. The thumb claw of *Iguanodon* is also famous. Once thought to have been a nose horn (see Chapter 2), it was a solid broad-based structure able to withstand a strong impact. It might have been used in fights between rival males, but it would also have been a useful defense against an attacking theropod.

## ● Sauropod thumb claws

It's unclear whether the sauropods were entirely immune from predation because of their large size or not. Despite the finds of theropod tooth marks on some skeletons (only very few), most paleontologists accept that, like modern elephants, the sauropods were pretty much left alone. Only the babies might be snatched. But sauropods did have some defensive equipment: most of them had a long claw on the thumb (see page 35), and this might have been flourished at predators.

## ● Titanosaur armor

The titanosaurs, an important group of Cretaceous sauropods, known mainly from South America but also from India, Madagascar, Africa, southern Europe, and southern North America, had armor. The armor plates were roughly round discs of bone set in the skin in regular close-packed patterns. It wasn't a solid armor like the shell of a turtle, but it formed something like chain mail, probably an effective defense against a passing attack. In a full-scale onslaught, however, a large theropod could bite past the armor, through to the vital organs, but would lose a few teeth in the process.

## ● The whiplash tail of *Diplodocus*

A rather startling suggestion was made in 1998: that a long-tailed sauropod like *Diplodocus* could crack its tail like a whip. Calculations by John Myrhvold, a computer engineer working for Microsoft, showed that *Diplodocus* could make the end of its tail travel at supersonic speed. If it had enough fast-action muscle power, a short sharp twitch at the root of its tail would translate into an enormous speed at the far-distant end. That is because a sideways twitch of three feet at the top of the tail is equivalent to a sideways movement of 50 to 65 feet at the far end. This is what gives the "crack" of the whip effect.
Critics have suggested that a *Diplodocus* might have been able to achieve such an effect, but only once. The supersonic speed would

fray the end of its tail, burst its eardrums, and give it such a sore tail and bottom that it would desist from such excesses forever.

### ● *Scelidosaurus* and the origins of armour

Fight and flight are all very well, but there is always a strong risk for the herbivore that it will be wounded or eaten nonetheless. Far better, perhaps, is a form of defense that offers one hundred percent protection. This was almost achieved by the armored dinosaurs, the stegosaurs, and especially the ankylosaurs. The first of these was *Scelidosaurus* from the Early Jurassic of southern England. *Scelidosaurus* did not have a full suit of armor, merely a series of rows of oval-shaped bone plates running along its back and sides and some short spines down the middle of its back. Not perfect, but a very good start.

### ● A good suit of armor

The armor suit was perfected by the ankylosaurs. The first ankylosaurs are reported from the Middle and Late Jurassic of Europe, and they had some more armor covering than did *Scelidosaurus*. But the apogee of ankylosaurian accoutrement came in the Late Cretaceous. Then, the famous ankylosaurs such as *Ankylosaurus*, *Euoplocephalus*, *Nodosaurus* and *Sauropelta* sported a complete suit of armor. The rows of bony plates over the back and sides had fused into a single solid armor coat. Some had long spines along the sides; some had tail clubs. Even their heads were completely encased in extra layers of bone.

### ● Ankylosaurs' armored eyelids

Ankylosaurs even had bony eyelid covers. Their armor plating was immensely thorough, with no opening left unprotected. If it were attacked, an ankylosaur presumably hunkered down to the ground, tucking in its arms and legs and ducking its head low. Unlike a turtle, though, it could not withdraw its head into its shell. So the ankylosaurs had a complex pavement of extra bones over the skull roof and around the sides of the snout. For safety, an ankylosaur could clamp its eyes tight shut. But eyes are attractive to predators, so the bony eyelid cover came into its own.

### ● Tail-whacking ankylosaurs

Ankylosaurs were not all passive when attacked. There were two groups of ankylosaurs, the nodosaurids and the ankylosaurids. Their key difference is that ankylosaurids had a tail club, and the nodosaurids did not. Perhaps the nodosaurids just had to sit tight while a hungry predator prowled round looking for a chink in the armor. An ankylosaurid, on the other hand, would take sly swipes with its bony tail club. The tail club was solid bone, produced from the fusion of the last few vertebrae in the tail. A good blow could stun a theropod or break its leg.

## ● The plates of *Stegosaurus*: upright or sideways?

The other armored dinosaurs, the stegosaurs, never achieved the solid armor covering seen in the ankylosaurs. Stegosaurs had rows of plates and spines down their back and additional spines on their shoulders and on the end of the tail. *Stegosaurus* itself, from the Late Jurassic of North America, famously had a row, or two rows, of large rhomboid plates set on end along its back. One paleontologist suggested that perhaps the plates lay flat, protecting the flanks, but their bone structure shows that they stood upright. Whether there was one row or two is still debated.

## ● Flushing to frighten

The plates of *Stegosaurus* might very well have had multiple functions. They could have acted as a radiator for shedding excess heat (see Chapter 6), they may have been part of a pre-mating ritual to frighten rival males, and they may have had a defensive purpose. Even though they did not protect the body, they may have been used to make *Stegosaurus* look bigger than it really was. In silhouette, with its plates, *Stegosaurus* looks twice its actual size. The plates were covered with skin, and they were well-supplied with blood vessels. Perhaps by flushing blood over the plates, *Stegosaurus* could increase the off-putting effect of fierceness and size.

## ● *Stegosaurus* tail spines

*Stegosaurus* had four long, sharp bone spines at the end of its tail. These were clearly for whacking and wounding predators. These round spines were typical of stegosaurs, and indeed they form the main armor in the stegosaurs from Africa and China. Here the principle might just have been to achieve random wounds by inducing the predator to lunge at its prey, and come up against a spike whenever it tried to get in close.

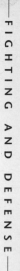

## ● Ornithopod defenses

Most ornithopods, all of them plant-eaters of course, were sleek runners, so speed was their main form of self-defense. But we've also seen that *Iguanodon*, for example, had a pretty strong, sharp thumb spike that may have been used in fights between males. But *Iguanodon* almost certainly wielded its claw against potential predators. Other ornithopods had minor defensive abilities: *Heterodontosaurus* from the Early Jurassic of South Africa had long canine teeth, and they all had strong flat-sided tails, especially the hadrosaurs. Once, the flat-sided hadrosaur tail was thought to have been a special adaptation for sustained swimming, but it could just as well have been used to deliver a stunning sideways blow.

## ● Defensive face horns

Ceratopsians, close relatives of the ornithopods, are distinguished by having horns on their faces (ceratopsian means "horned face") and a bony frill at the back of the skull. The oldest ceratopsian, *Psittacosaurus* from the Mid Cretaceous of Mongolia, was a biped, not so very different from *Iguanodon*, but with the thickened snout, downturned "beak" and a small frill. Later ceratopsians all had a nose horn, immensely long in some such as *Monoclonius*. Others had at least a pair of horns above the eye sockets, as in *Triceratops*, while yet others had as many as five face horns (*Pentaceratops*) and some, like *Styracosaurus*, had wicked horns around the neck frill. These were all good for seeing off predators. Faced by a row of ceratopsians looking straight at it, the theropod would have had to retire in confusion.

## ● The ceratopsian frill

What was the neck frill for? In early ceratopsians, the frill was little more than a short shelf projecting behind the skull, but in later forms this shelf extended for some six to ten feet. It probably had a defensive function in protecting the neck, but it must also have been used for display, threat and other purposes. Protecting the neck is all very well, but a frill is a complicated way to do so, and the rest of the body behind the neck was still vulnerable to attack.

CHAPTER NINE

# DINOSAUR EVOLUTION AND BIRDS

Throughout the book, we've been flinging around terms with abandon such as "theropod," "sauropod," "stegosaur" and the like. It's time now to establish clearly just what these groups were. Do they relate to evolution or are they just convenient categories based on general shape and behavior? In recent years there has been a vigorous discussion about the pattern of dinosaurian evolution, and huge strides have been made. If you compare a dinosaur book of ten or fifteen years ago with the present view, you can see enormous changes in our views of the origin and evolution of dinosaurs. The debate continues.

## ● What is a dinosaur?

Dinosaurs are not a random assortment of giant and horrible extinct animals. The group Dinosauria is a real evolutionary unit. This means that all dinosaurs evolved from a single common ancestor, and that it is possible to trace the exact pattern of evolution within the group. The family tree of dinosaurs is reconstructed by studying their anatomy and by searching for unusual features that may be shared by two or more species. Close study of the details of the dinosaurian family tree in the past twenty years has shown that dinosaurs originated from a single ancestor 230 to 235 million years ago, that they branched out early on into many lines, and that they live on today in the form of birds.

## ● Debating the dinosaurian debut

When Sir Richard Owen named the Dinosauria in 1842 (see Chapters 2 and 3), he understood that they had had a single ancestor. But, in later Victorian times, and through much of the twentieth century his insight was denied. Paleontologists argued that the dinosaurs had arisen from two, three, even five separate ancestors. If this had been true, then the "Dinosauria" wouldn't be a real evolutionary group. But new work since 1980 has shown that dinosaurs, and no other reptiles, share a number of unique characteristics, and this new work proves their origin from a single ancestor.

## ● Defining the Dinosauria

Dinosaurs stood fully upright on their hind legs, which were tucked right under their bodies (see Chapter 5). This upright posture caused major changes in the orientation of the various leg joints: there are more than two sacral vertebrae (the vertebrae that fix the hip joint to the backbone); the hip joint has a deepened socket for the ball-like head of the femur; the knee and ankle are simple back-and-forward hinges; the ankle bones are reduced to narrow rollerlike structures that fit on to the end of the shinbones (especially the tibia); the foot is digitigrade (the dinosaur stands on tiptoe). If you find a reptile with these features, it's a dinosaur. But where did they come from?

## ● The monster success of the dinosaurs

In evolutionary terms, the dinosaurs were hugely successful. As a group, they dominated the earth for 165 million years, living on all continents and evolving into many different lines. They include small(ish), medium, large, and very large forms, including some of the largest land animals ever. They also held back the evolution of the mammals. Humans are mammals, and we think that mammals, with their intelligence, parental care, warm-bloodedness, and so on, are the pinnacle of evolutionary achievement. But the first mammals appeared about 225 million years ago, at the same time as the first dinosaurs. And the dinosaurs clearly dominated the mammals then.

## ● Dinosaurs are ruling reptiles

In terms of their place in the scheme of things, dinosaurs are archosaurs. The Archosauria ("ruling reptiles") include today crocodiles and birds, but in the past the group included also the pterosaurs, a range of basal forms in the Triassic, and, of course, the dinosaurs. So the dinosaur evolved from basal archosaurs of the Middle Triassic, presumably from some kind of smallish bipedal insect-eating archosaur.

## ● A dinosaur ancestor

The most likely ancestors of dinosaurs are the lagosuchids, known from the end of the Middle Triassic, perhaps 230 to 235 million years ago, of South America. *Marasuchus*, the best-known lagosuchid was a lightly built flesh-eater, some four feet long. It presumably preyed on small fast-moving animals, early lizardlike reptiles, as well as perhaps worms, grubs, and insects. The skeleton shows many dinosaur-like characters, such as a swanlike S-curved neck, an arm that is less than half the length of the leg, and the beginnings of an opening in the hip joint. *Marasuchus* was clearly a biped, running on its hind limbs, and the long tail was presumably used as a balancing organ. It may have used its hands for grappling with prey and for passing food to its mouth.

## ● The oldest dinosaur?

There has been a long-running debate about the oldest dinosaur remains. Earlier accounts frequently stated that the dinosaurs arose early in the Triassic, and they often quoted evidence in the form of skeletons and footprints. However, the supposed skeletal remains of dinosaurs from before the Late Triassic turn out to belong to a variety of non-dinosaurian groups. Dinosaur footprints, which have a characteristic three-toed appearance (see Chapter 5), had also been recorded from the Early and Mid Triassic of various parts of the world, but critical re-examination has now shown that they have been wrongly identified.

## ● Dinosaurs of the Ischigualasto Formation

The oldest true dinosaurs are known from the early part of the Late Triassic. The best specimens come from the Ischigualasto Formation of the Ischigualasto region of Argentina (Ischigualasto means "valley of the moon"). The Ischigualasto dinosaurs, *Eoraptor* and *Herrerasaurus*, are relatively well known from nearly complete specimens (see page 143), and they give an insight into the days before the dinosaurs rose to prominence. *Eoraptor* and *Herrerasaurus* were minor components of their faunas, so these first dinosaurs apparently did not come to dominate the scene. More on this later.

## ● First finds of *Herrerasaurus*

The bones of slender bipedal dinosaurs were found in the Ischigualasto Formation of Argentina in the 1950s and 1960s and named *Herrerasaurus*. This animal was identified right away as a dinosaur: it had

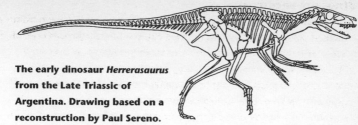

**The early dinosaur *Herrerasaurus*
from the Late Triassic of
Argentina. Drawing based on a
reconstruction by Paul Sereno.**

long slender hind limbs that were held in an upright posture, the
hipbones had a perforation where the head of the femur fitted in a
ball-and-socket joint, the knee and ankle joints formed simple hinges,
and the foot was arranged so that the animal stood high on its toes,
the digitigrade posture.

## New finds

A new wave of collecting in the Ischigualasto Formation in the 1990s
turned up a spectacular new skeleton of ten to twenty feet in length.
There is a dispute, however, about how much of a dinosaur
*Herrerasaurus* truly is: is it more primitive than all subsequent dinosaurs,
or can it be assigned to the Theropoda, the group of carnivorous forms?
Paul Sereno of Chicago has championed the idea that *Herrerasaurus* is a
theropod based on its possession of a specialized joint in the lower jaw,
seen only in those meat-eating dinosaurs. But other dinosaur
paleontologists have been unable to see this claimed joint, and they
argue that *Herrerasaurus* is a dinosaur, but not a theropod.

## *Eoraptor*, the most primitive dinosaur

Paul Sereno and his Argentinian colleague, Eduardo Novas, made
another spectacular discovery during their collecting in the 1990s.
They unearthed the skeleton of another, perhaps more primitive,
dinosaur, which they named *Eoraptor*, or "dawn hunter." *Eoraptor* was
much smaller than *Herrerasaurus*, being about three feet long. It had
all the dinosaur characteristics seen in *Herrerasaurus*, including a strong
three-fingered hand. However, the skull of *Eoraptor* represents a basic
dinosaur design, with no specializations hinting at Theropoda, or
any of the other major dinosaur groups.

## ● Biology of the first dinosaurs

*Herrerasaurus* and *Eoraptor* were agile hunters that could run fast and maneuvre effectively. They had the advantage of speed and of having arms that were free for grabbing and grappling. They could prey on the smaller animals of Ischigualasto times. Top carnivores were still the rauisuchians, a group of basal archosaurs. New studies of the ecology of Ischigualasto times show that these early dinosaurs were a minor part of their communities, representing perhaps five percent of all animals. The age of the dinosaurs had begun, but in rather a discreet manner!

## ● Competition or mass extinction?

There are currently two ways of viewing the radiation of the the dinosaurs in the Late Triassic. Either they "took their chance" after a mass extinction event and radiated opportunistically or they competed over a longer time span with the preexisting reptiles and eventually prevailed. Landscapes of the earlier Triassic were dominated by three main reptile groups. The basal archosaurs included a range of mainly carnivorous forms, most of them quadrupedal hunters, but some, like the lagosuchid *Marasuchus*, were fleet little bipeds. The second group were the rhynchosaurs, strange bulky herbivores with beaked heads and rows of crushing teeth. The third group were the mammal-like reptiles, plant- and flesh-eaters, that included the ancestors of the mammals.

## ● The classic model for dinosaurian origins

Until the 1980s most paleontologists assumed that the competitive model was correct for three reasons. First, the origin of the dinosaurs was seen as a drawn-out affair, that started well down in the Middle Triassic and ran on for tens of millions of years. Secondly, the appearance of the dinosaurs has often been regarded as a great leap forward in evolutionary terms, and so there must have been some purpose to it. This leads to the third reason: a general view that evolution is progressive, and so major replacements like this must involve the evident superiority of the new group. The superiority of the first dinosaurs was in their erect gait, or their supposed warm-bloodedness, or their sharper teeth, or something.

## ● Fossil evidence for dinosaurian origins

The pattern in the fossil record does not support the competitive model. Dinosaurs appeared at about the beginning of the Late Triassic, 230 million years ago, and they were rare elements in their faunas (less than five percent of individuals) for the next 10 million years. Dinosaurs diversified only about 220 million years ago, and all the major lineages appeared during this time. After 220 million years ago, dinosaurs represent fifty to ninety percent of individuals, clear ecological dominance, and a dramatic increase from their previous totals.

## ● An end-Carnian extinction event?

There was evidently a major extinction event about 220 million years ago at the end of the Carnian stage in the late Triassic. Various families of basal archosaurs, mammal-like reptiles, and the rhynchosaurs died out. The key extinctions were those of the dominant herbivore groups, the plant-eating mammal-like reptiles and the rhynchosaurs. The disappearance of these medium-sized and large plant-eaters would be like the loss of all deer, cattle, and elephants from the African plains today. There was a huge ecological void. The cause of this extinction event is unclear. There was a major shift in paleoclimates at the time, from generally wet to much drier, and plants changed also. The *Dicroidium* flora (low, leafy, seed ferns) of the southern hemisphere gave way to a worldwide conifer flora at about this time. Perhaps the change in plant food was enough to wipe out the herbivores.

## ● Did the dinosaurs have superior adaptations?

The "superior adaptations" of dinosaurs were probably not so profound as was once thought. Many other archosaurs also evolved erect gait in the Late Triassic, and yet they died out. The physiological characteristics of dinosaurs – whether they were warm-blooded or not, for example – cannot be determined with confidence (see Chapter 6). This is not to say that the first dinosaurs were not well adapted and efficient predators. Just that it is not clear that they were so much better than all the other Triassic reptiles.

## ● Ecological replacements: selection or good luck?

The idea that competition can have major long-term effects in evolution is probably an oversimplification of a complex set of processes. Obviously, competition happens in nature. Darwin himself realized that competition was critical in evolution. Male animals compete for mates. There is competition between different species, perhaps if they are feeding on the same limited food, or if one is hunting the other. But the concept of one whole evolutionary group of plants or animals competing with another major group is a much less obvious concept. Perhaps dinosaurs owed their original success to good luck. After all, such a model is quite familiar. It is pretty well accepted that the mammals radiated 65 million years ago only after the bit of good luck (for them) that the dinosaurs were wiped out by some other cause.

## ● How many dinosaurs are known?

Dinosaurs were vastly successful, of that there is little doubt. But just how successful? So far, about 1500 dinosaur species have been named, but many of these are not entirely convincing. Some were based on incomplete specimens, sometimes just an odd bone or tooth, and others turned out to be the same as species that had already been named. So, for example, the well-known sauropod *Brontosaurus* was named in 1878, but it was noticed later that it was the same as *Apatosaurus*, which had been named in 1877. The name *Apatosaurus* has precedence, and it must be used in preference to the later name *Brontosaurus*. The true number of named dinosaur species is much smaller than 1500 because of this synonymy problem, perhaps 800 species at present.

## ● How many dinosaurs actually existed?

If no one was finding any new dinosaurs, we might assume that the file was closed. The final tally is 1500 (or better 800) species. But new species are being reported from all over the world, and it seems that they are being discovered at an increasing rate! While only one or two species were announced each year in the 1960s, this has risen to the startling figure of twenty new species in 1999 alone. Perhaps the total will keep on rising. But, sadly, people do not learn from past mistakes, so it is likely that at least some of these twenty 1999 dinosaurs are synonymous with previously named forms. At the past rate of something like forty percent synonymy, the total comes down to twelve truly new dinosaurs named in 1999. Since I was a contributor to one of those new names in 1999, I'd better watch out!

## ● Regional trends in discovery

If dozens of new dinosaur species are still turning up, then it might seem impossible to say anything about the actual total number that really existed. But if the figures are examined continent by continent, something interesting emerges. Most of the new species are coming from virgin parts of the world, areas that have not been heavily explored for dinosaurs. Admittedly, a few of the new species come from Europe and North America, but not many, since these continents have been thoroughly picked over since early in the nineteenth century. But paleontologists have collected seriously in Asia and the southern continents only from 1900 onwards, so many treasures are still to be found in these largely unexamined areas.

## ● Sorting out the classification of the Dinosauria

How are the dinosaurs to be divided up into subgroups? The Dinosauria is a natural evolutionary group, just like birds, mammals, or insects. It can be defined by a number of unique characteristics, just as feathers and wings define birds and hair and milk production define mammals, for example. The key subdivisions of the dinosaurs are, broadly speaking, pretty obvious. Everyone over the age of two can explain the difference between a theropod and a sauropod, a stegosaur and an ankylosaur, a ceratopsian, and a pachycephalosaur.

## ● First steps: the Saurischia and the Ornithischia

The first level of division of Dinosauria is into the two major subgroups Saurischia and Ornithischia. In 1887 the English paleontologist Harry Seeley noted that dinosaurs fell into two great groups based on their pelvic structures, the Saurischia ("reptile hips") and Ornithischia ("bird hips"). Saurischians show the classic reptilian hip structure, with the two lower hipbones, the pubis, and ischium, pointing respectively forwards and backwards (see below). Ornithischians, on the other hand, have a

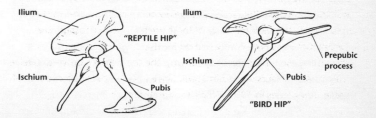

Ilium     Ilium

"REPTILE HIP"     Prepubic process

Ischium     Ischium     Pubis

Ischium     Pubis

"BIRD HIP"

**The hips of saurischian (left) and ornithischian (right) dinosaurs, showing the different orientations of the pubis.**

modified hip structure, in which the pubis has swung backwards to run parallel with the ischium.

# Dividing the dinosaurs further

The Saurischia are divided into Theropoda, the meat-eating dinosaurs, and Sauropodomorpha, the large long-necked plant-eaters. Ornithischia are also subdivided into various subgroups: the two-legged Ornithopoda, the horn-faced Ceratopsia, the plate-spined Stegosauria, and the armored Ankylosauria (see page 150). These subdivisions became quite clear after the work of Cope and Marsh in the latter part of the nineteenth century (see Chapter 3). But the subdivisions also caused some confusion, since paleontologists believed that some of the subgroups of dinosaurs had had independent origins, as many as four or five.

# A revolution in systematics

Debates in systematics had gone on for a long time, and not just concerning dinosaurs. Systematics is the study of the relationships and evolution of life. Part of systematics is the drawing up of evolutionary trees, and the attempt to make them as close to the truth as possible. A key advance came in the 1960s with the adoption of a new method of tree reconstruction called cladistics. Cladistics was developed by a German entomologist, Willi Hennig. His insight was that, in classification, one should use only derived (advanced) characteristics. So, in sorting out the pattern of dinosaurian evolution, look only for those characteristics that are unique to specific subgroups.

# Cladistics in action

There is often some confusion about the difference between derived characteristics (which are useful in systematics) and primitive characteristics (which are not). Dinosaurs offer a good example. Cladistic analyses have confirmed the existence of the Saurischia and the Ornithischia. The Saurischia are defined, not by their possession of a "lizardlike" pelvis, since that feature is primitive, shared with all other reptiles, including, of course, lizards. Saurischian characters include specific features of the skull, backbone, and hand. The Ornithischia, on the other hand, *are* defined by their unique pelvic arrangement, since it is seen in no other animals.

● **The computer revolution**

At first, dinosaur paleontologists made their evolutionary tree reconstructions by hand. They made up great tables of all the characters that were present or absent in dozens of dinosaur species and tried to find the best-fitting tree to link them together by hand. This is a laborious process, and it doesn't always achieve the best result. After 1985, computer programs became widely available for tree-building calculations. This advance has revolutionized the practice of cladistics, and every dinosaurologist now feeds screeds of data into his computer and churns out evolutionary trees all over the place. This is not to say they are all right, but at least the trees can be cross-checked and tested by others.

● **Building the dinosaur family tree**

Cladistics confirmed the broad patterns that had already been established (see page 150). Within Saurischia, the classic division into Theropoda and Sauropodomorpha was confirmed, and the Ornithischia were also found to divide up into the traditional groups, the Ornithopoda, Ceratopsia, Stegosauria ,and Ankylosauria. So far so good. But could cladistics offer anything new?

● **New insights from cladistics**

The new cladistic analyses of Dinosauria, carried out after 1985, did clarify a number of previously contentious issues. An additional ornithischian group, the Pachycephalosauria, bipedal herbivores with massively thickened skull roofs, was recognized. Cladistic analysis showed that the Ceratopsia and Pachycephalosauria shared a close relationship, and they are amalgamated in the Marginocephalia (see page 150). The marginocephalians and ornithopods then pair off as the Cerapoda. The armored dinosaurs, the stegosaurs, and ankylosaurs, together form the group Thyreophora, and Thyreophora plus Cerapoda make up the Ornithischia.

● **Dinosaur groups come and go**

It is now possible to survey briefly the coming and going of the different dinosaur lines through the 165 million years of their existence. The first thing that is clear is that the dominance of different groups changed through that long span of time. A simple view, that somehow all the well-known dinosaurs were trotting about together at the same time,

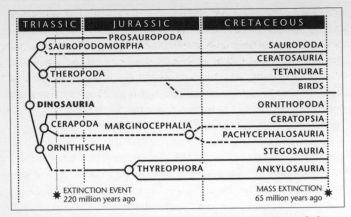

**Evolutionary tree of the dinosaurs, showing the main groups, their relationships and their distribution in time.**

is clearly nonsensical. For a start, no species of dinosaur lasted for more than five million years, so such a view is clearly impossible.

## ● Early splitting in the dinosaurian family tree

A great deal of splitting evolution happened among dinosaurs in the Late Triassic. The oldest dinosaurs, *Eoraptor* and *Herrerasaurus*, may be basal theropods, or perhaps even generalized basal dinosaurs. But whatever they are it is clear that the main lines of the dinosaur family tree (see above) had already split early in the Late Triassic. The Saurischia/Ornithischia split happened before the extinction event 220 million years ago. Also, the Theropoda/Sauropodomorpha split happened in this time.

## ● The Late Triassic: first dinosaur faunas

Typical Late Triassic dinosaurian faunas consisted of prosauropods, plant-eaters such as *Plateosaurus* from the Late Triassic of Germany. The theropods in the Late Triassic did not reach large size. *Coelophysis* from Arizona and New Mexico, ten feet long, is a typical example. Ornithischians were rare small bipedal forms.

## ● Dinosaurs of the Early Jurassic

Early Jurassic dinosaur faunas were very similar to those of the Late Triassic. They were dominated by prosauropods as herbivores, and

*Coelophysis*-like theropods as carnivores. Towards the end of the Early Jurassic, the first sauropods, descended from the prosauropods, came on the scene. These early sauropods, represented by isolated finds from South Africa, India, and Germany, showed key changes in skull pattern (larger eyes, nostrils shifted back, teeth restricted to the front of the jaw) and in the skeleton (more vertebrae in the neck, reinforced limb girdles, pillarlike limbs, simplified foot skeletons). This was the beginning of the rise of the monster dinosaurs.

## ● Sauropoda

The sauropods were the biggest land animals of all time. Some of them, such as *Brachiosaurus* reached lengths of 76 feet or more and heights of 43 feet. The sauropods were most abundant during the Jurassic, and they became much less significant during the Cretaceous, although some sauropod groups, especially the titanosaurs, survived until the very end of that period. In their heyday, the sauropods could be diverse. The Morrison Formation, of Late Jurassic age, has produced many skeletons of six or seven different sauropods, all apparently living contentedly side by side (see Chapter 4).

## ● Radiation of the ceratosaurian theropods

The Theropoda, the flesh-eating dinosaurs, radiated during the Jurassic. One group, the ceratosaurs, animals such as *Coelophysis* from the Late Triassic, had branched off early on. The ceratosaurs diversified into larger forms in the Jurassic. Some, like *Dilophosaurus* from the Early Jurassic of North America, and *Ceratosaurus* from the Late Jurassic of Tanzania and North America, had cranial excrescences of one sort or another: parallel platelike crests in the former, low horns in the latter. These may have been used in sexual displays.

## ● The tetanurans

A new theropod group, the tetanurans, rose to prominence in the Jurassic. These were mainly larger theropods, animals such as *Megalosaurus* and *Gasosaurus* from the Middle Jurassic and *Allosaurus* from the Late Jurassic. The tetanurans had a distinguished evolutionary record, giving rise also to both the birds in the Jurassic (more of this later) and an astonishing array of theropods both small and large in the Cretaceous.

## ● Thyreophora, the armored dinosaurs

Stegosaurs and ankylosaurs form together the Thyreophora. The first thyreophorans appeared in the Early Jurassic, animals like *Scelidosaurus* in England and *Scutellosaurus* in North America. The group split in the Middle Jurassic into the stegosaurs and ankylosaurs, but the latter group is known in the Jurassic only from limited remains, and radiated mainly in the Late Cretaceous. The stegosaurs, on the other hand, are best represented in the Middle Jurassic of China and the Late Jurassic of Tanzania and the American Midwest. *Stegosaurus*, from the Late Jurassic of North America, had a row of bony plates along the middle of its back, this feature may have had a temperature-control function (see Chapter 6).

## ● Ornithopoda: amazing tooth batteries

The ornithopod dinosaurs are known first in the Late Triassic and Early Jurassic from a number of small rare forms. By Late Jurassic times, ornithopods such as *Camptosaurus* had achieved moderately large size, and they were coming into their own in a landscape still dominated by sauropods. The sauropods declined after the end of the Jurassic, and the ornithopods became dominant as herbivores in the Early Cretaceous. They owed their success, in all probability, to their massive batteries of teeth, which allowed effective browsing on tough vegetation. In addition, they were all bipeds, and they specialized in speed of movement to escape from larger predators.

## ● Cretaceous flesh-eating dinosaurs

The two divisions of Theropoda, the ceratosaurs and the tetanurans, had different fates in the Cretaceous. The ceratosaurs continued as the abelisaurids, a modest-sized group known best from southern continents. The tetanurans, on the other hand, blossomed and flourished. From ancestors like *Megalosaurus* and *Allosaurus* in the Jurassic, tetanurans small and large evolved in the Cretaceous.

## ● Tetanuran diversity

Cretaceous tetanurans included a huge array of predators, from the modest-sized *Deinonychus* to the huge *Tyrannosaurus*. The dromaeosaurids, like *Deinonychus* and *Velociraptor*, are famous for their slashing claws (see Chapter 8). Their relatives, the troodontids, had a similar claw, and they were distinguished by extra-large brains. Two groups of toothless tetanurans were important in the Late Cretaceous, *Oviraptor* and its relatives in Mongolia (see Chapters 7 and 8) and the ornithomimosaurs, the ostrich dinosaurs with their light, slender frames and potential for high speed (see Chapter 5). The tyrannosaurids, *Tyrannosaurus* and its relatives from the Late Cretaceous of North America and Central Asia, were huge flesh-eaters, perhaps relatively slow-moving when compared to their smaller brethren.

## ● Success of the ornithopods

The main plant-eating dinosaurs of the Cretaceous were the ornithopods. Their tooth batteries were well adapted for dealing with the new, fast-growing flowering plants. In the Early Cretaceous, the hypsilophodontids lived nearly worldwide, from the Wealden of England (*Hypsilophodon*) to the South Pole regions of Australia (*Leaellynasaura*). Iguanodontids were another highly successful group in Europe (*Iguanodon*) and Africa (*Ouranosaurus*).

## ● The hadrosaurs: sheep of the Mesozoic

The most diverse, and most succesful,ornithopod clade were the hadrosaurs or "duck-billed" dinosaurs of the Late Cretaceous. Frequently, hundreds of individuals of several distinct hadrosaurian species are found side by side in the same geological formation. Huge herds roamed the plains of North America and central Asia. It's no wonder they have been called the "sheep of the Mesozoic". The hadrosaurs are famous for their expanded ducklike bills and their massive dental batteries, which allowed them to process tough plant food (see Chapter 4). They are also well known for their amazing array of crests (see Chapter 7)

## ● The pachycephalosaurs

The pachycephalosaurs were a small group of mainly Late Cretaceous herbivores from North America and central Asia. They are well known for their remarkably thick skull roofs which may, or may not, have been used in head butting contests (see Chapter 8). The pachycephalosaurs are close relatives of the ceratopsians.

## ● The ceratopsians

The Ceratopsia (literally "horned faces") comprise about twenty genera known mainly from the Late Cretaceous of North America. All are characterized by a triangular-shaped skull when viewed from above, an additional beaklike rostral bone in the midline at the tip of the snout, a high snout and an expanded frill over the neck region. *Protoceratops* from the Late Cretaceous of Mongolia and China was a small form which had a thickened bump in front of its eyes. Later relatives of the ceratopsians had elaborate frills and horns, which probably had both defensive and display functions.

# Birds as living dinosaurs

Thomas Henry Huxley got it right in 1870 when he pointed out that birds are just small flying dinosaurs. He had been impressed by the skeleton of *Archaeopteryx*, found in 1861. Stripped of its feathers, he noticed, it was just a small theropod dinosaur. With its feathers it was a bird. Despite this, it took another hundred years of bickering before people began to see that Huxley's observation was correct.

# *Archaeopteryx*

*Archaeopteryx* is probably the most famous single fossil in the world. Ever since the discovery of the first example in 1861, this fossil bird has attracted huge attention. It was hailed by the evolutionists as the ideal "missing link," or proof of evolution in action. Here was an animal with a beak, wings, and feathers, so it was clearly a bird, but it still had a reptilian bony tail, claws on the hand and teeth. Since 1861, six more skeletons have come to light, the last two in 1987 and 1992.

# The biology of *Archaeopteryx*

*Archaeopteryx* was about the size of a magpie, and it fed on insects. The claws on its feet and hands suggest that *Archaeopteryx* could climb trees, and the wings are clearly those of an active flying animal. This bird could fly as well as most modern birds, and flying allowed it to catch prey that were not available to land-living relatives.

# Alternative views about bird origins

The bird-dinosaur hypothesis was not popular until about 1970. During the early twentieth century, many ideas were suggested. Most popular was the theory that birds had originated directly from basal archosaurs of the Triassic, but others have suggested an origin from Late Triassic crocodilian relatives, or from a joint ancestor with mammals. These views held sway for decades, even though there were no fossils of the

**Reconstruction of *Caudipteryx*, a dinosaur from the Early Cretaceous of China, announced in 1998. This is a troodontid-like dinosaur, but with feathers preserved.**

supposed archosaur-bird intermediates during the long time gap from Late Triassic to latest Jurassic, about fifty million years.

## ● The tide turns

John Ostrom's work on the small raptor *Deinonychus* turned the tide. The skeleton of *Archaeopteryx* is very like that of *Deinonychus*, especially in the details of the arm and hind limb, showing that birds are small flying tetanuran theropod dinosaurs. Opponents of this view have hunted high and low for pre-birds in the Triassic/Jurassic interval, so far without success. They still cling to their belief that birds are not dinosaurs, but their case was lost in 1970.

## ● Bird evolution after *Archaeopteryx*

Until 1990 very little was known about bird evolution during the bulk of the Cretaceous. Othniel Marsh had described in the 1880s some remarkable toothed birds from the Late Cretaceous of North America, *Hesperornis* and *Ichthyornis*. Only odd scraps of other birds had come to light before the appearance of modern birds at the very end of the Mesozoic, sixty-five million years ago. Then, some astonishing new specimens were announced, first four or five specimens from Spain, and then dozens from China (see above). These new specimens were all exquisitely well preserved, showing every bone, feathers, even beaks and claws. They allowed paleo-ornithologists to fill lots of gaps in the evolutionary tree of birds. Next time you see a sparrow twittering in the hedgerow, remember you are looking a cousin of *Tyrannosaurus* in the eye. And be very, very scared.

## CHAPTER TEN

# EXTINCTION

The extinction of the dinosaurs exerts an endless fascination. It's often the only question that people ask: "Well, why did they die out?" But the extinction of the dinosaurs is not such an unusual phenomenon. The fate of all species is to die out sooner or later. Species are not eternal: they last typically from two to ten million years, during which time they either evolve into something else or disappear. The dinosaur extinction debate has been a major discussion point for an astonishing diversity of scientists since 1980. It has become a dynamic, "big science" and heavily funded area of research. The ongoing debate is also a fascinating microcosm of the way in which scientists work. I'll show the evidence for the different viewpoints, but also try to lift the lid on exactly what's going on behind some of the postures and polemics.

## ● Ongoing extinction

The popular question about the extinction of the dinosaurs refers to their final demise, 65 million years ago. Of course all 800, or 1500, species of dinosaurs that ever existed (see Chapter 9) did not all run up to the line and then keel over. There had been continuing evolution and extinction of dinosaurs throughout the Mesozoic. As we've seen in Chapter 9, the key forms came and went. Major groups, too, had their time on earth and were replaced by others.

## ● Theories about extinction

Ever since the first discovery of dinosaurs, scientists have speculated about their extinction. Richard Owen was the first. In 1842, in the same paper in which he established the name "Dinosauria," he also discussed why they might have disappeared. He suggested that, being reptiles of a sort, they must have had lower oxygen requirements than mammals. Therefore, oxygen levels in the Mesozoic must have been lower than they are today. At the end of the Cretaceous, he argued, oxygen levels rose to modern levels, and mammals and birds took over. The dinosaurs turned up their toes. As a style of explanation, this stinks. It's perfect circular reasoning: if A, then B, and if B, then A is proved.

## ● Racial senility

Early in the twentieth century, the most popular notion about dinosaur extinction was racial senility. In a nutshell, this idea was that the dinosaurs were exhausted, finished, clapped out, mere remnants of their former selves, and that they became extinct simply because they had served their time. Specific evidence was found in the fact that many Late Cretaceous dinosaurs, such as the ceratopsians and hadrosaurs, had horns and crests on their heads. These were said to show wild patterns of evolution that merely produced useless structures that encumbered the animals. With such extraordinary and purposeless cranial excrescences, these animals were clearly on the way out.

## ● Dinosaur = overblown and inaccurate

Even though ideas of racial senility have long been rejected, the word "dinosaur" is still used in that context. The image of dinosaurs as inevitable failures lives on with many people who still use the term "dinosaur" to mean anything that is too big and inefficient to survive. This is an unsustainable view, since dinosaurs were clearly some of the most successful animals ever to have lived. The horns and crests of Late Cretaceous dinosaurs were not useless, and there is no evidence for racial senility as the cause of the extinction.

## ● Wild and preposterous notions

I have counted more than a hundred specific theories for the extinction of the dinosaurs that have appeared between 1842 and 1970. They were too big, too stupid, too undersexed, too ill. Fred Hoyle even suggested they died of AIDS brought down to earth by cosmic impacts. Experts on paleoclimate claimed that it became too hot or too cold, too wet or too dry, too seasonal, or that the atmosphere became too poor in oxygen, or full of poison gases: take your pick. Biologists claimed that the dinosaurs were wiped out by competition: that the mammals

ate all their food, or at least ate their eggs, that caterpillars ate all the
plants, that the tyrannosaurids were so voracious that they ate every last
herbivore, and then keeled over themselves. Wild-eyed visionaries
claimed that the earth had been blasted by solar winds, cosmic rays,
sunspot activity, or that it had been hit by meteorites, asteroids, comets,
and other forms of extraterrestrial debris.

## ● Unfettered speculation

It seems amazing now to look back at this phase of wild speculation.
There is probably no other area in science where otherwise intelligent
people feel they are entirely free to speculate in such an unfettered way.
I might think, for example, that the sure-fire cure for cancer is to stand
on your head and consume leeks. But I wouldn't get very far if I tried to
publish the idea in a scientific journal. The quality control procedures –
the checking of manuscripts by knowledgeable experts – would put a
stop to that immediately. But dinosaur extinction was different:
anything went (and still goes, I'm afraid to say).

## ● Untestable nonsense

All the older speculative notions about dinosaur extinction are
unsatisfactory because they are untestable. Owen's oxygen levels
argument goes on circling round and round, and some independent
evidence is needed to break the loop. "Racial senility" said nothing, just
that the dinosaurs died out because they were preordained to die out.
Most of the other theories fail since there is actually no evidence for
them and, more important, no possibility of testing (and rejecting)
them. A scientific theory is a theory that can be rejected.

## ● The last dinosaurs: on their way out?

The last dinosaurs were some of the most varied ever seen. Dominant
by far were the hadrosaurs, or duck-billed dinosaurs. A visitor to the Late
Cretaceous of North America, or indeed Mongolia or China, would have
seen vast herds of hadrosaurs, ankylosaurs, ceratopsians, and various
theropods. Latest Cretaceous theropods include the small, highly
intelligent troodontids, the larger ostrichlike ornithomimids, and the
largest carnivores of all, the tyrannosaurids. Bones of *Triceratops* and
*Tyrannosaurus* are found within a few yards of the last sediments of the
Cretaceous in Montana, so these highly successful dinosaurs were some
of the last to walk the earth.

# ● What else died out?

Many groups of plants and animals other than the dinosaurs died out 65 million years ago. Among terrestrial tetrapods, the pterosaurs also disappeared, as well as several families of birds and marsupial mammals. In the sea, plesiosaurs, mosasaurs, and a few families of sharks and fish disappeared. Many other important Mesozoic groups popped their clogs at about the same time, such as the ammonites, belemnites, rudist and trigoniid bivalves, and various plankton groups in the sea. Now, bearing all this in mind, it is clear why many of the older theories of dinosaur extinction collapse: they say nothing about the loss of all these other groups.

# ● What survived?

Most plants and many animals, however, were apparently unaffected by the cataclysms 65 million years ago. Survivors include the gastropods (snails, whelks, and relatives), most bivalves (mussels, clams, and relatives), fish, amphibians, turtles, lizards, and placental mammals. It's just as important to look at what survived, as what went extinct, when trying to understand a mass extinction. The fact that so many groups survived proves that whatever happened cannot have been so ghastly that everything in sight was killed instantly.

# ● Was the extinction selective?

It is hard to separate the survivors and non-survivors into simple ecological categories. It would not be correct to say, for example, that only big animals (like the dinosaurs) died out, since some birds and small marine organisms disappeared too. Paleontologists have made many detailed studies in which they have compared survivors and victims of the mass extinction 65 million years ago. They have found only one factor that may have been important – not size, diet, habitat, or anything like that. Species that belonged to geographically widely distributed groups seemed to survive better than locally distributed forms. That seems to make some sense.

# ● The KT event

The mass extinction that marked the end of the dinosaurs is always called the KT event. So we have to get used to that. The letters "K" and "T" stand for the two geological periods involved, the Cretaceous and the Tertiary. Well, of course Cretaceous begins with a "C," but the letter C is already used as the abbreviation for the older Carboniferous period. So geologists use "K" for Cretaceous, rationalizing that "K" stands for *kreta*, "chalk" in Greek, the origin of the name Cretaceous.

# ● Mass extinctions

The KT event obviously was one of a number of mass extinctions. Paleontologists talk about extinction, extinction events, and mass extinctions. This is a ranking in order of severity. Extinction can refer to the loss of even a single species – the extinction of the dodo for example – or to the loss of a smaller or larger group. An extinction event is a restricted time-interval during which a broadish range of species disappeared. An example is the extinction event at the end of the great Ice Ages, 10,000 years ago, when mammoths, mastodons, and woolly rhinos died out. More serious events are called mass extinctions.

# ● A definition of mass extinctions

So what is a mass extinction? A precise numerical definition cannot be given. A mass extinction is an event during which large numbers of species of diverse ecology disappeared in a short interval of time. "Large numbers" means perhaps twenty percent of species or more worldwide. "Diverse ecology" means plants and animals, dwellers on land and sea, organisms large and small. The loss of a few large hairy mammals, as at the end of the Ice Ages, does not make a mass extinction. A "short interval of time" means short in geological terms, so certainly less than a million years. In a sense, having been around for 165 million years, a short time for the extinction of the dinosaurs could be as much as 5 million years. Or as little as a day. More of this later.

# ● Rates of extinction

Paleontologists like to talk about high and low rates of extinction. There are many different ways of assessing rates. Some of them give a measure of risk, or probability, of extinction. The simplest measure is the number of species (or groups) going extinct, divided by the amount of time (in millions of years). Another simple measure is the percentage loss. So, extinction rates for the KT event are a hundred percent loss for dinosaurs and zero percent loss for lizards. If, for example, there were only ten species of dinosaurs around in the last million years of the Cretaceous, then their loss represents an extinction rate of ten species per million years.

## ● The "big five" mass extinctions

There were five mass extinctions in the past 500 million years. Before that, it is hard to be sure, because the quality of the fossil record tails off rapidly in such ancient rocks. The so-called "big five" events were these:

1) Late Ordovician (450 million years ago)
2) Late Devonian (370 million years ago)
3) End Permian (250 million years ago)
4) End Triassic (205 million years ago)
5) End Cretaceous (KT; 65 million years ago).

## ● Are we living through the sixth mass extinction?

Experts often suggest that we are currently living through the sixth mass extinction. Certainly, if you figure out the rates of extinction caused by human activities, they add up to high figures. Among birds, for example, humans have probably caused the extinction of 100 species in the past 400 years, in other words an extinction rate of 25 per 100 years, or 250 per 1000 years, or 250,000 per 1,000,000 years. If birds truly continued to be wiped out at this rate, of course, all 7000 living species would be gone in 2800 years. These rates are comparable to those of the "big five" mass extinctions in the past.

## ● How rapid was it?

It is impossible for anyone to say exactly how long the KT event took. There are two levels of difficulty. First, radiometric dating techniques (not radiocarbon dating, since that cannot measure dates older than about 40,000 years, but methods like it) cannot be used on most rocks, and even where they can be used there are errors of up to five percent. Five percent of 65 million years is three and a quarter million years. The latest methods give errors of less than one percent, but that's still plus or minus 650,000 years. The second problem concerns the patterns of decline shown by the fossils. How accurate are they?

## ● The Signor-Lipps effect

Some parts of the fossil record of the last parts of the Cretaceous seem to show a gradual decline, perhaps over one or two million years. As paleontologists collected up through the rock sequence, species seemed to drop out every few yards until only one or two were left immediately below the KT boundary. But paleontologists realized that this might not be the true pattern. It's pretty clear that not every plant and

animal that ever lived will be preserved as a fossil. So in collecting,
however carefully, it's unlikely that the last fossil of a species is actually
the last of that species that ever lived. This is the Signor-Lipps effect,
named after the paleontologists who pointed it out. A pattern of
seeming decline might just represent the dropping out of the last fossils,
but not a true pattern of decline.

## ● Patterns of extinction

Did all the plant and animal groups that died out do so essentially at the
same time (catastrophic event) or over a span of several million years
(gradual event)? Allowing for the Signor-Lipps effect, the evidence
suggests that some groups disappeared catastrophically right at the KT
boundary while others were in long decline before the end of the
Cretaceous. The ichthyosaurs, for example, disappeared thirty million
years earlier than the KT boundary, while the belemnites, ammonites,
plesiosaurs, and pterosaurs had dwindled to low diversities and had
become quite rare.

## ● A new era of research on the KT event

Up to 1980 something like five or six papers were published each year
about the KT event. Most of these were rather vague, often crazy, ideas
about why the dinosaurs died out. This desultory interest was shot
through as if by a bolt of lightning in 1980. Since then, some 500 to
1000 papers each year have been devoted to the subject. This has been
one of the most amazing revolutions in earth sciences in living memory.

## ● Professor Alvarez's bombshell

The intense interest in the KT event was galvanized by a single ten-page
paper published in 1980. In this paper, Luis Alvarez, of the University of
California at Berkeley, and colleagues published their view that the
extinctions had been caused by the impact of a huge meteorite, six
miles across, on the earth. The impact caused massive extinctions by
throwing up a vast dust cloud that blocked out the sun and prevented
photosynthesis: hence, plants died off, followed by herbivores
and then carnivores.

## ● The paleontologists are provoked

Luis Alvarez's 1980 paper could not be ignored. Unlike many of the
writers on the KT event up to 1980, Alvarez was a highly distinguished

scientist. He had twice won the Nobel Prize for Physics. He wrote forcefully, and his argument seemed to be watertight (we'll look at the evidence in a minute). But most paleontologists were appalled. How dare a physicist jump into their domain and try to hijack it? And how dare he make such wild claims? Impacts and explosions? This, they thought, was the stuff of fantasy.

## ● Boldness = good science

What Alvarez had done in his 1980 paper was hugely daring, and hence the best kind of science. His proposal was so detailed and so dramatic that it could have been disproved readily by all kinds of evidence. Remember, science is about probabilities, not certainties. Except in mathematics, you can't prove a theory, you can only try to reject it. The more a theory stands up to the assault of attempts at disproof, the stronger it becomes. Alvarez has been assaulted repeatedly since 1980 (metaphorically speaking), and his ideas have stood firm.

## ● Not only bold, but heuristic

The Alvarez paper was especially daring because the authors really had very little evidence for their wild hypothesis. The heuristic quality of a scientific hypothesis is the scope of its predictiveness, how much can be found out as a result of an idea ("heuristic" comes from the Greek for "to find," and it's related to Archimedes' famous cry of "Eureka,—I found it!"). So the Alvarez hypothesis was vastly heuristic, especially in comparison with the small body of evidence on which it was based. The idea opened up a huge array of new research fields, from paleontology to cosmology, climatic models to disaster modeling.

## ● Nuclear winter

This heuristic quality became evident immediately. Disaster modelers who were trying to figure out the potential effects of an all-out nuclear war realized that this kind of artificial disaster had much in common with the KT natural disaster. A nuclear explosion throws up huge clouds of dust and gas and creates a vast crater. They realized that a vast dust cloud encircling the globe would cut out not only light but also heat from the sun. And so was born the concept of "nuclear winter," the freezing episode after a catastrophe. Another heuristic effect was the proposal, in 1983, of repeated, predictable mass extinctions. We'll come back to those.

## ● The evidence

The Alvarez team had really just one piece of evidence for their astonishingly daring hypothesis for the KT event. They had been working on a rock section in Gubbio, in northern Italy, looking for the element iridium. Iridium is a platinum-group element that is rare on the earth's crust, and reaches the earth from space in meteorites. The Alvarez team was hoping to use the iridium to assess rates of deposition of the rocks, but they were shocked to find a huge leap in the quantity of iridium in a thin clay layer right bang on the KT boundary. This could mean one of two things: a sudden huge influx of iridium, or that the thin clay layer represented a huge span of time and the iridium had become concentrated.

## ● The iridium spike

The Alvarez team interpreted the elevated levels of iridium, called an "iridium spike," in the KT boundary beds as evidence of a huge increase in the rate of accumulation of that element. In other words, an impact. They confirmed the iridium spike at another locality, in Denmark, and then wrote their paper. Although the iridium spike shows a multiplication of the amount of iridium by ten times or more, note that the actual quantities are tiny. Iridium is measured in *parts per billion*, vanishingly minute quantities. It was to Luis Alvarez's credit that he had devised a way to record these minute quantities, something that had been quite impossible until the late 1970s.

## ● Retrodicting from the iridium spike

The whole model for the impact scenario was drawn up by working backwards from the iridium spike. This could be called retrodiction, predicting in a backwards way. The reasoning went like this: iridium spike in two localities equals iridium spread around at least the northern hemisphere (if not the whole world), which means a globe-encircling dust cloud, which means an impact powerful enough to throw up enough material to create such a cloud, which means a crater 60 to 120 miles across, which means a giant rock (asteroid/meteorite) six miles across. And that was that.

## ● New evidence

While many people criticized the Alvarez theory, and rejected it as wildly speculative, other scientists set about pursuing confirming evidence.

Geologists scoured KT locations around the world. Between 1980 and 2000, some 200 such sites were found to show iridium spikes. In other words, pretty well everywhere you looked, if there was a KT boundary layer clay, you'd find an iridium spike in it. That was an amazing confirmation of Alvarez's leap of faith.

## ● Shocked quartz and glassy spherules

In the 1980s two new pieces of physical evidence came to light. Some KT boundary clays contained crystals of quartz, a common mineral in rocks. But these quartz crystals were shocked quartz, i.e. quartz with numerous multiple sets of parallel planes through the crystal structure. It is well known that such parallel lamellae, as they're called, are produced under high pressure. They can be found after some explosive volcanic eruptions, but these are less well developed than the KT examples. Also, many of the boundary clays contained small glass beads, melt glasses, too. Such melt glasses can indeed come from volcanoes, but the chemical composition of most of them was quite unlike any volcanic product and much more related to melt materials from under an impact crater.

## ● The fern spike

A catastrophic extinction is indicated by abrupt shifts in pollen ratios at some KT boundaries. Pollen is shed by a variety of plants, and is incorporated into sediments, often in huge quantities. It provides a good way to follow the relative success of different plant types round a deposition site. The shifts in pollen ratios show a sudden loss of flowering plants and their replacement by ferns, and then a progressive return to normal floras. This fern spike, found at many terrestrial KT boundary sections, is interpreted as indicating the aftermath of a catastrophic ash fall: ferns recover first and colonize the new surface, followed eventually by the flowering plants after soils begin to develop. This interpretation has been made by looking at what happens today after major volcanic eruptions.

## ● Killing dinosaurs

The impact model for the KT extinction event might seem to be so strong that nothing would have survived. So, say the critics, did the impact actually kill the dinosaurs? These critics point out, quite reasonably, that it is all very well finding evidence for an impact, and tracing its effects worldwide, but, they ask, how did it actually do the killing, and why was the kiling selective? Remember that a broad range of animals on land and in the sea were killed off, but many others were hardly affected.

## ● Gradual extinction of the dinosaurs?

There is certainly a great deal of evidence that the dinosaurs were *not* wiped out overnight. Dinosaurs, and many other groups of organisms appear to have been in decline long before the KT boundary and the impact. So how to explain this? Either the fossil record is hopelessly inaccurate, and the Signor-Lipps effect (see above) is a serious problem, or there truly was a decline. The record of dinosaurs does not simply show a gradual decline through the last few yards of the Cretaceous, but over the last several hundred yards, possibly the last five million years. Or, then again, does it?

## ● Collecting thousands of dinosaur specimens

In the early 1990s two teams decided to comb the Hell Creek Formation in Montana with fine-tooth combs. The idea was to try to resolve once and for all exactly what had happened to the dinosaurs. Had they declined gradually over five million years or was it an instant kill-off, right at the KT boundary? The two teams, using slightly different approaches, looked inch by inch through the thickness of the Hell Creek Formation, documenting thousands of fossil occurrences bed by bed. The result? As you've guessed, one team came to the convincing conclusion that dinosaurs were declining step by step over the last five million years of the Cretaceous, and the other team found evidence of an instant die-off. The teams continue to bicker in the pages of the scientific journals.

## ● The gradualist model

The gradual dying off of the dinosaurs is easy to understand in terms of long-term climatic changes. It's not clear whether this story is true worldwide. All the work so far has been done in North America. Subtropical lush dinosaurian habitats gave way to strongly seasonal, temperate, conifer-dominated mammalian habitats. There is little doubt about the climatic deterioration in North America in the closing millions of years of the Cretaceous. Seems reasonable? But how can you measure ancient climates?

## ● Deteriorating climates

The deteriorating climates in the latest Cretaceous of North America are
documented by two main methods. First, paleobotanists see major
changes in temperature-sensitive plants. Modern plants often have very
specific temperature and moisture preferences. Some of the Late
Cretaceous plants are close relatives of modern forms, and so these
climatic factors can be determined quite precisely. The measurements
are confirmed by physical measurements using oxygen isotopes. These
climatic changes on land are linked to changes in sea level and in the
area of warm shallow water seas. So the gradualists can explain all the
extinctions by their earth-bound model, and without any input from
the impact theory.

## ● The Deccan Traps: major volcanic eruptions

Another twist is that most of the KT phenomena may be explained by
volcanic activity. The Deccan Traps in India represent a vast outpouring
of lava which occurred over the million years or so spanning the KT
boundary. Supporters of the volcanic model seek to explain all the
physical indicators of catastrophe (iridium, shocked quartz, spherules,
and the like) and the biological consequences as the result of the
eruption of the Deccan Traps. The volcanic model fits quite
comfortably with the evidence of longer-term decline in certain
groups, but it also ties in with sudden extinctions when eruptive
activity peaked.

## ● Periodic mass extinctions

An astonishing claim was made by two paleontologists from Chicago
in 1983. David Raup and Jack Sepkoski suggested that the mass
extinctions had been periodic, that they were exactly spaced through
time. If true, this claim meant that we could predict precisely when the
next one would happen. They believed they had found evidence that
mass extinctions in the past 250 million years happened regularly every
26 million years. The last one happened 16 million years ago. The cause
of the regular cycle, or periodicity, was said to be astronomical.
Astronomers set about looking for freak phenomena that would
regularly shower the Earth with meteorites every 26 million years. The
debate became pretty quiet after the 1980s since the postulated
astronomical agents were not found. Also, repeated re-study of the
fossil records did not strongly support the idea of regular mass

extinctions. But if Raup and Sepkoski were right, it's one of the most amazing discoveries, and all an outcome of the original 1980 paper by Luis Alvarez and his team.

## ● Pinpointing the crater

One of the weaknesses of the Alvarez impact model was that there was no smoking gun. Where was the crater? At first, supporters of Alvarez claimed that the crater had been buried under later-deposited rocks, or that the crater was beneath the waves, or that it had been eaten up at a volcanically active area: for example, around Iceland. In fact, a crater wasn't needed for Alvarez to be right, but, equally, a crater would do much to complete the story.

## ● The crater is discovered

After a number of false starts, discovery of the crater was announced in 1990. Earlier announcements of the crater in Iowa and off Colombia were quickly forgotten. It was pinned down, lying half-onshore, half-offshore, on the Yucatán Peninsula in southern Mexico. Smack in the middle of the crater lies the village of Chicxulub (see page 169), so it is now called the Chicxulub crater, much to the bemusement of the startled locals.

## ● Dr. Hildebrand's detective work

The crater wasn't located by chance, and nor could it be seen. It was found after some dogged detective work by Alan Hildebrand, a Canadian graduate student. People were already homing in on the Caribbean. The KT boundary layers were thicker there than anywhere else. On the Caribbean island of Haiti, in one section, the KT layer is 30 inches thick, rather than the usual half an inch. On the east coast of Mexico, and in southern Texas, boundary layers were sometimes several yards thick. A gradient of thickness leads you to the culprit, because any fallout deposits are thickest at the source, tapering out as you move away.

## ● Tsunamites

A new kind of rock deposit, called a tsunamite, also pointed towards the middle of the Caribbean. In the 1980s geologists had identified beds of apparently tumbled and overturned rocks at several locations in Texas and Mexico. The tumbled blocks were beach rocks, often quite large,

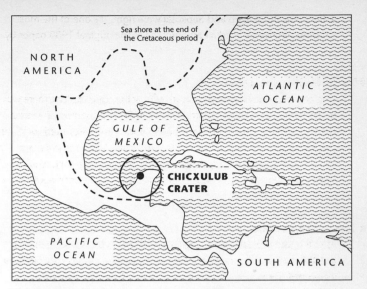

**The site of the KT crater, centred on the village of Chicxulub, on the Yucatán Peninsula of Mexico. The ancient Proto-Caribbean coastline is shown: it was hit by a massive tsunami soon after the impact.**

indicating some powerful force that had hit the ancient coastline. The geologists immediately thought of massive tidal waves, commonly called tsunamis, their name in Japanese. Rock deposits created by tsunamis are called, of course, tsunamites. But what on earth had been powerful enough to throw up tsunamites all around the fringes of the Caribbean?

## ● Impact in the sea

The Chicxulub crater, now half on land and half under the sea, was located entirely within the Late Cretaceous proto-Caribbean sea. So the six mile asteroid had plunged into the sea and had drilled the huge crater into the marine rocks of the ancient sea floor. The tsunamis had been set up by that impact, and they then headed out at top speed in all directions from the impact site. Those tsunamis would have been hundreds of yards high at the source, and they traveled as fast as express trains. As they shot sideways, friction slowed both their speed and cut them down in height. But their effect on the coastlines only a few hundred miles away would have been devastating. But the tsunamis, however, weren't the key killer (although any dinosaurs sunbathing within 60 miles of the beach zone would have been killed).

# ● Airborne debris

The first effect of the impact was to throw a huge mass of material into the sky. To every action there is an equal and opposite reaction. If you punch a hole into the Earth, with something like a thousand times the energy of all the nuclear arsenals on earth today, there has to be a backlash. Millions of tons of sea floor were thrown high in the air, and these blocks rained down for hundreds of miles around the crater site. But that wasn't all.

# ● The dust cloud

Much more significant was the dust cloud. The meteorite itself, all its millions of tons, would have vaporized on impact. This vaporized material, plus the fine-scale debris from the sea bed, and melted and blasted rock fragments, were all sent into the sky. The finer the particles, the higher they go. The gas and dust shot into the upper airless atmosphere and formed a vast plume. Depending on the wind direction, it would have blown across North America, North Africa, or Europe (remember that the Atlantic Ocean was much narrower then). The dust cloud would easily have encircled the whole of the northern hemisphere, if not the whole of the earth. And it would take years for all the dust to disperse and settle out.

# ● Wildfire

Many KT boundary beds have produced soot particles, blackened blobs and fluffy bits. The soot is found especially in North America, so perhaps the wind was blowing north-westwards at the time. It's believed that after the physical backlash to the impact, the throwing out of debris, there was a chemical effect. The huge amounts of energy, interacting with chemicals in the bedrock, sent out explosive gases, and a great weighty curtain of burning matter marched out over the landscape. This set fire to all the forests it encountered, and these probably burned for weeks, long after the hot gas cloud had dispersed.

# ● Killing dinosaurs

So is that enough to kill every dinosaur in sight or not? Certainly in the zone lying hundreds of miles downwind of the impact site, nothing would have been left living. Total devastation. But we must keep reminding ourselves that the mammals, birds, lizards, snakes, turtles, plants, and most of the life in the water were not really affected, or at least not in the long term. Key studies now will aim to see whether there were geographical areas where life was blitzed, and areas where life was unaffected.

## ● The aftermath

After the impact, certainly North America and some other parts of the world would have presented a pretty bleak picture. But life has astonishing powers of recovery. Within days of the inferno, ferns would uncurl from beneath the ash. Some pond creatures would have survived. Insects seem to be able to survive anything. Evidently, the smaller vertebrates were able to shelter somehow. But the dinosaurs had finally been removed from the face of the earth. As is well known, the mammals and birds took over. Where dinosaurs had once walked across the plains, mammals took their place, although it took them ten million years to achieve large size and real ecological diversity.

## ● Consensus

Amazingly, though, for such an astonishing event, which had such long-term ramifications, we still know so little. Available killing models are either biologically unlikely, or too catastrophic: remember that a killing scenario must take into account of the fact that 75 percent of families survived the KT event, many of them seemingly entirely unaffected. Whether the two models – impact and long-term climatic deterioration – can be combined so that the long-term declines are explained by gradual changes in sea level and climate and the final disappearances at the KT boundary were the result of impact-induced stresses is hard to tell. And then there were those massive eruptions in India...

APPENDIX

# THE DINOSAUR GENERA

This list gives all 796 genera (plural of genus) of dinosaurs named up to 2000. It's the first time such a list has been given in a general book, and at this time of writing it is comprehensive and complete – I hope. Most of these genera contain only a single species, but many have more than one, so a full list of species would run to 1500 or so. There seems to be no sign of a slowing-down in the rate of naming new dinosaurs: some twenty new genera were named in 1999, and we've even sneaked in six new forms from 2000 – see if you can find them. Some of the genera turn out not to be genuinely new; these are noted and their correct assignment given. I have not included genera that have at some time or another been called dinosaurian, but are now known not to be.

The total of 796 genera is probably far lower than the actual total of different dinosaur genera that ever lived – new forms are being found all the time. However, the figure of 796 is an overestimate of the true number of genera that is known; many of the named genera are now known to be synonyms, the same as genera that have already been named, others are *nomina nuda* or *nomina dubia*. These technical terms should be explained. *Nomen dubium* (' dubious name') is a name given to a scrappy or very incomplete specimen which is quite inadequate to support a new name; *nomen nudum* ('naked name') is a name that has been published but is not supported by an adequate description or illustration.

In the present list, there are 158 synonymous genera, 87 *nomina nuda*, and 127 *nomina dubia*, bringing the grand total of named and accepted dinosaur genera to 424. This is calculated from 796 minus the 'error' names (158 + 87 + 127 = 372). This is a remarkably high rate of error, some 372 of 796 named forms (forty-seven per cent). Maybe dinosaur palaeontologists have always been too quick to stick a name on a new bone or tooth, or maybe they are just far too critical, and quick to find fault in the work of their colleagues.

# FASCINATING FACTS

## A

| | | | |
|---|---|---|---|
| *Abelisaurus* Bonaparte and Novas, 1985 | Argentina | Late Cretaceous | Theropoda: Abelisauridae |
| *Abrictosaurus* Hopson, 1975 | South Africa | Early Jurassic | Ornithopoda: Heterodontosauridae |
| *Abrosaurus* Ouyang, 1989 (*nomen nudum*) | | | |
| *Acanthopholis* Huxley, 1867 | UK | Late Cretaceous | Ankylosauria: Nodosauridae |
| *Achelousaurus* Sampson, 1995 | USA | Late Cretaceous | Ceratopsia: Ceratopsidae |
| *Achillobator* Perle, Norell, and Clark, 1999 | Mongolia | Late Cretaceous | Theropoda: Dromaeosauridae |
| *Acrocanthosaurus* Stovall and Langston, 1940 | USA | Early Cretaceous | Theropoda: Acrocanthosauridae |
| *Acrocanthus* Langston, 1947 (*Acrocanthosaurus*) | | | |
| *Adasaurus* Barsbold, 1983 | Mongolia | Late Cretaceous | Theropoda: Dromaeosauridae |
| *Aegyptosaurus* Stromer, 1932 | Egypt | Late Cretaceous | Sauropoda: Titanosauridae |
| *Aeolosaurus* Powell, 1987 | Argentina | Late Cretaceous | Sauropoda: Titanosauridae |
| *Aepisaurus* Gervais, 1853 (*nomen dubium*) | | | |
| *Aetonyx* Broom, 1911 (*Massospondylus*) | | | |
| *Afrovenator* Sereno and others, 1994 | Niger | Early Cretaceous | Theropoda: Torvosauridae? |
| *Agathaumas* Cope, 1874 (*Edmontosaurus*) | | | |
| *Agilisaurus* Peng, 1990 | China | Mid-Jurassic | Ornithopoda: Fabrosauridae? |
| *Agrosaurus* Seeley, 1891 (*Thecodontosaurus*) | | | |
| *Agustinia* Bonaparte, 1999 | Argentina | Early Cretaceous | Sauropoda: Titanosauridae? |
| *Alamosaurus* Gilmore, 1922 | USA | Late Cretaceous | Sauropoda: Titanosauridae |
| *Albertosaurus* Osborn, 1905 | Canada/USA | Late Cretaceous | Theropoda: Tyrannosauridae |
| *Albisaurus* Fritsch, 1905 (*nomen dubium*) | | | |
| *Alectrosaurus* Gilmore, 1933 | Mongolia | Late Cretaceous | Theropoda: Tyrannosauridae |
| *Algoasaurus* Broom, 1904 (*nomen dubium*) | | | |
| *Alioramus* Kurzanov, 1976 | Mongolia | Late Cretaceous | Theropoda: Tyrannosauridae |
| *Aliwalia* Galton, 1985 | South Africa | Late Triassic | Theropoda: family? |
| *Allosaurus* Marsh, 1877 | USA | Late Jurassic | Theropoda: Allosauridae |
| *Alocodon* Thulborn, 1973 (*nomen dubium*) | | | |
| *Altirhinus* Norman, 1998 | Russia | Late Cretaceous | Ornithopoda: Hadrosauridae |
| *Altispinax* Huene, 1923 (*Megalosaurus*) | | | |
| *Alvarezsaurus* Bonaparte, 1991 | Argentina | Late Cretaceous | Theropoda: Alvarezsauridae |
| *Alwalkeria* Chatterjee and Creisler, 1994 | India | Late Triassic | Theropoda: family? |
| *Alxasaurus* Russell and Dong, 1994 | China | Late Cretaceous | Theropoda: Therizinosauridae |
| *Amargasaurus* Salgado and Bonaparte, 1991 | Argentina | Early Cretaceous | Sauropoda: Diplodocideae |
| *Ammosaurus* Marsh, 1891 | USA | Early Jurassic | Prosauropoda: Plateosauridae? |
| *Ampelosaurus* Le Loeuff, 1995 | France | Late Cretaceous | Sauropoda: Titanosauridae |
| *Amphicoelias* Cope, 1877 (*nomen dubium*) | | | |
| *Amphisaurus* Marsh, 1882 (*Anchisaurus*) | | | |
| *Amtosaurus* Kurzanov and Tumanova, 1978 (*nomen dubium*) | | | |
| *Amygdalodon* Cabrera, 1947 | Argentina | Mid-Jurassic | Sauropoda: Cetiosauridae |
| *Anasazisaurus* Hunt and Lucas, 1993 | USA | Late Cretaceous | Ornithopoda: Hadrosauridae |
| *Anatosaurus* Lull and Wright, 1942 (*Edmontosaurus*) | | | |
| *Anatotitan* Chapman and Brett-Surman, 1990 | USA | Late Cretaceous | Ornithopoda: Hadrosauridae |
| *Anchiceratops* Brown, 1914 | USA | Late Cretaceous | Ceratopsia: Ceratopsidae |
| *Anchisaurus* Marsh, 1885 | USA | Early Jurassic | Prosauropoda: Anchisauridae |
| *Andesaurus* Calvo and Bonaparte, 1991 | Argentina | Early Cretaceous | Sauropoda: Titanosauridae |
| *Angaturama* Kellner and Campos, 1996 | Brazil | Early Cretaceous | Theropoda: Spinosauridae |
| *Animantarx* Carpenter and others, 1999 | USA | Late Cretaceous | Ankylosauria: Nodosauridae |
| *Ankylosaurus* Brown, 1908 | USA/Canada | Late Cretaceous | Ankylosauria: Ankylosauridae |
| *Anodontosaurus* Sternberg, 1929 (*Euoplocephalus*) | | | |
| *Anoplosaurus* Seeley, 1878 (*nomen dubium*) | | | |
| *Anserimimus* Barsbold, 1988 | Mongolia | Late Cretaceous | Theropoda: Ornithomimidae |
| *Antarctosaurus* Huene, 1929 | Argentina | Late Cretaceous | Sauropoda: Titanosauridae |
| *Anthodon* Owen, 1876 (*Paranthodon*) | | | |
| *Antrodemus* Leidy, 1870 (*Allosaurus*) | | | |
| *Apatodon* Marsh, 1877 (*Allosaurus*) | | | |
| *Apatosaurus* Marsh, 1877 | USA | Late Jurassic | Sauropoda: Camarasauridae |
| *Aragosaurus* Sanz and others, 1987 | Spain | Early Cretaceous | Sauropoda: Camarasauridae |
| *Aralosaurus* Rozhdestvenskii, 1968 | Russia | Late Cretaceous | Ornithopoda: Hadrosauridae |
| *Archaeoceratops* Dong and Azuma, 1997 | China | Late Cretaceous | Ceratopsia: Ceratopsidae |
| *Archaeornithoides* Elzanowski and Wellnhofer, 1992 | Mongolia | Late Cretaceous | Theropoda: family? |
| *Archaeornithomimus* Russell, 1972 | China | Late Cretaceous | Theropoda: Ornithomimidae |
| *Arctosaurus* Adams, 1875 (*nomen dubium*) | | | |
| *Argentinosaurus* Bonaparte and Coria, 1993 | Argentina | Mid-Cretaceous | Sauropoda: Titanosauridae |
| *Argyrosaurus* Lydekker, 1893 | Argentina | Late Cretaceous | Sauropoda: Titanosauridae |
| *Aristosaurus* Van Hoepen, 1920 (*Massospondylus*) | | | |
| *Aristosuchus* Seeley, 1887 (*Calamospondylus?*) | | | |

*Arkansaurus* Sattler, 1983 (*nomen nudum*)

| | | | |
|---|---|---|---|
| *Arrhinoceratops* Parks, 1927 | Canada | Late Cretaceous | Ceratopsia: Ceratopsidae |

*Arstanosaurus* Suslov and Shilin, 1982 (*nomen dubium*)

| | | | |
|---|---|---|---|
| *Asiaceratops* Nesov and Kaznyshkina, 1989 | Uzbekistan | Early Cretaceous | Ceratopsia: Protoceratopsidae |

*Asiatosaurus* Osborn, 1924 (*nomen dubium*)

*Astrodon* Johnston, 1859 (*Pleurocoelus?*)

*Astrodonius* Kuhn, 1961 (*Pleurocoelus?*)

*Atlantosaurus* Marsh, 1877 (*Apatosaurus*)

| | | | |
|---|---|---|---|
| *Atlasaurus* Monbaron, Russell, and Taquet, 1999 | Morocco | Mid-Jurassic | Sauropoda: family? |
| *Atlascopcosaurus* Rich and Rich, 1989 | Australia | Early Cretaceous | Ornithopoda: Hypsilophodontidae |

*Aublysodon* Leidy, 1868 (*nomen dubium*)

*Augustia* Bonaparte, 1998 (*Agustinia*)

*Austrosaurus* Longman, 1933 (*nomen dubium*)

| | | | |
|---|---|---|---|
| *Avaceratops* Dodson, 1986 | USA | Late Cretaceous | Ceratopsia: Ceratopsidae |

*Avalonia* Seeley, 1898 (*Camelotia*)

| | | | |
|---|---|---|---|
| *Avimimus* Kurzanov, 1981 | Mongolia | Late Cretaceous | Theropoda: Avimimidae |

*Avipes* Huene, 1932 (*nomen dubium*)

*Azendohsaurus* Dutuit, 1972 (*nomen dubium*)

# B

| | | | |
|---|---|---|---|
| *Bactrosaurus* Gilmore, 1933 | USA | Late Cretaceous | Ornithopoda: Hadrosauridae |
| *Bagaceratops* Maryanska and Osmólska, 1975 | Mongolia | Late Cretaceous | Ceratopsia: Ceratopsidae |
| *Bagaraatan* Osmólska, 1996 | Mongolia | Late Cretaceous | Theropoda: family? |

*Bahariasaurus* Stromer, 1934 (*nomen dubium*)

| | | | |
|---|---|---|---|
| *Bambiraptor* Burnham and others, 2000 | USA | Late Cretaceous | Theropoda: Dromaeosauridae? |
| *Barapasaurus* Jain and others, 1975 | India | Early Jurassic | Sauropoda: family? |
| *Barosaurus* Marsh, 1890 | USA | Late Jurassic | Sauropoda: Diplodocidae |
| *Barsboldia* Maryanska and Osmólska, 1981 | Mongolia | Late Cretaceous | Ornithopoda: Hadrosauridae |
| *Baryonyx* Charig and Milner, 1987 | UK | Early Cretaceous | Theropoda: Spinosauridae |

*Bashunosaurus* Kuang, 1996 (*nomen nudum*)

*Basutodon* Huene, 1932 (*Euskelosaurus*)

| | | | |
|---|---|---|---|
| *Becklespinax* Olshevsky, 1991 | UK | Early Cretaceous | Theropoda: family? |

*Beelemodon* Bakker, 1997 (*nomen nudum*)

| | | | |
|---|---|---|---|
| *Beipiaosaurus* Xu, Tnag and Wang, 1999 | China | Early Cretaceous | Theropoda: Therizinosauridae |
| *Bellusaurus* Dong, 1990 | China | Late Jurassic | Sauropoda: Titanosauridae? |

*Betasuchus* Huene, 1932 (*nomen dubium*)

*Bibleyhallorum* Burge and others, 1999 (*nomen nudum*)

*Bihariosaurus* Marinescu, 1989 (*nomen dubium*)

| | | | |
|---|---|---|---|
| *Blikanasaurus* Galton and Van Heerden, 1985 | South Africa | Late Triassic | Sauropoda: family? |
| *Borogovia* Osmólska, 1987 | Mongolia | Late Triassic | Theropoda: Troodontidae |

*Bothriospondylus* Owen, 1875 (*nomen dubium*)

| | | | |
|---|---|---|---|
| *Brachiosaurus* Riggs, 1903 | USA/Tanzania | Late Jurassic | Sauropoda: Brachiosauridae |
| *Brachyceratops* Gilmore, 1914 | USA | Late Cretaceous | Ceratopsia: Ceratopsidae |
| *Brachylophosaurus* Sternberg, 1953 | USA | Late Cretaceous | Ornithopoda: Hadrosauridae |

*Brachypodosaurus* Chakravarti, 1934 (*nomen dubium*)

*Brachyrophus* Cope, 1878 (*nomen dubium*)

| | | | |
|---|---|---|---|
| *Breviceratops* Kurzanov, 1990 | Mongolia | Late Cretaceous | Ceratopsia: Protoceratopsidae |

*Brontosaurus* Marsh, 1879 (*Apatosaurus*)

| | | | |
|---|---|---|---|
| *Bruhathkayosaurus* Yadagiri & Ayyasami, 1989 | India | Late Cretaceous | Theropoda: Tetanurae? |
| *Bugenasaura* Galton, 1995 | USA | Late Cretaceous | Ornithopoda: Hypsilophodontidae |
| *Byronosaurus* Norell, Makovicky & Clark, 2000 | Mongolia | Late Cretaceous | Theropoda: Troodontidae |

# C

| | | | |
|---|---|---|---|
| *Caenagnathasia* Currie, Godfrey & Nesov, 1994 | Mongolia | Late Cretaceous | Theropoda: Oviraptoridae |

*Caenagnathus* R. M. Sternberg, 1940 (*Chirostenotes*)

| | | | |
|---|---|---|---|
| *Calamosaurus* Lydekker, 1891 (*?nomen dubium*) | | | |
| *Calamospondylus* Fox, 1866 | England | Early Cretaceous | Theropoda: Coelurosauria? |

*Calamospondylus* Lydekker, 1889 (*Calamosaurus*)

*Callovosaurus* Galton, 1980 (*Camptosaurus*)

| | | | |
|---|---|---|---|
| *Camarasaurus* Cope, 1877 | USA | Late Jurassic | Sauropoda: Camarasauridae |
| *Camelotia* Galton, 1985 | England | Late Triassic | Prosauropoda: Melanorosauridae |
| *Camposaurus* Hunt and others, 1998 | USA | Late Triassic | Theropoda: Coelophysidae |

*Camptonotus* Marsh, 1879 (*Camptosaurus*)

| | | | |
|---|---|---|---|
| *Camptosaurus* Marsh, 1885 | USA | Late Jurassic | Ornithopoda: Camptosauridae |

*Campylodon* Huene, 1929 (*Campylodoniscus*)

*Campylodoniscus* Kuhn, 1961 (*nomen dubium*)

| | | | |
|---|---|---|---|
| *Capitalsaurus* Kranz, 1998 (*nomen nudum*) | | | |
| *Carcharodontosaurus* Stromer, 1931 | Egypt | Mid-Cretaceous | Theropoda: Allosauridae |
| *Cardiodon* Owen, 1841 (*nomen dubium*) | | | |
| *Carnosaurus* Huene, 1929 (*nomen nudum*) | | | |
| *Carnotaurus* Bonaparte, 1985 | Argentina | Mid-Cretaceous | Theropoda: Abelisauridae |
| *Caseosaurus* Hunt and others, 1998 | USA | Late Triassic | Theropoda: Herrerasauridae? |
| *Cathetosaurus* Jensen, 1988 (*Camarasaurus*) | | | |
| *Caudipteryx* Ji Q., Currie, Norell & Ji S., 1998 | China | Early Cretaceous | Theropoda: Maniraptora |
| *Caudocoelus* Huene, 1932 (*nomen dubium*) | | | |
| *Caulodon* Cope, 1877 (*Camarasaurus*) | | | |
| *Cedarosaurus* Tidwell, Carpenter & Brooks, 1999 | USA | Early Cretaceous | Sauropoda: Titanosauridae |
| *Centrosaurus* Lambe, 1904 | Canada | Late Cretaceous | Ceratopsia: Ceratopsidae |
| *Ceratops* Marsh, 1888 (*nomen dubium*) | | | |
| *Ceratosaurus* Marsh, 1884 | USA | Late Jurassic | Theropoda: Ceratosauridae |
| *Cetiosauriscus* Huene, 1927 | England | Late Jurassic | Sauropoda: Neosauropoda? |
| *Cetiosaurus* Owen, 1841 | England | Mid-Jurassic | Sauropoda: Cetiosauridae |
| *Changtusaurus* Zhao, 1983 (*nomen nudum*) | | | |
| *Chaoyangsaurus* Zhao, Cheng & Xu, 1999 | China | Late Cretaceous | Ceratopsia: Ceratopsidae |
| *Chasmosaurus* Lambe, 1914 | Canada | Late Cretaceous | Ceratopsia: Ceratopsidae |
| *Chassternbergia* Bakker, 1988 (*Edmontonia*) | | | |
| *Cheneosaurus* Lambe, 1917 (*Hypacrosaurus*) | | | |
| *Chialingosaurus* Young, 1959 | China | Mid-Jurassic | Stegosauria: Stegosauridae |
| *Chiayusaurus* Bohlin, 1953 (*Asiatosaurus*) | | | |
| *Chihuahuasaurus* Ratkevich,1997 (*nomen nudum*) | | | |
| *Chilantaisaurus* Hu, 1964 | China | Early Cretaceous | Theropoda: Carnosauria? |
| *Chindesaurus* Long & Murry, 1995 | USA | Late Triassic | Theropoda: Herrerasauridae? |
| *Chingkankousaurus* Young, 1958 | China | Late Cretaceous | Theropoda: Neotetanurae |
| *Chinshakiangosaurus* Yeh, 1975 (*nomen nudum*) | | | |
| *Chirostenotes* Gilmore, 1924 | Canada | Late Cretaceous | Theropoda: Elmisauridae |
| *Chondrosteosaurus* Owen, 1876 | England | Early Cretaceous | Sauropoda: Camarasauridae? |
| *Chondrosteus* Owen, 1876 (*Chondrosteosaurus*) | | | |
| *Chuandongocoelurus* He, 1984 (*nomen nudum*) | | | |
| *Chubutisaurus* del Corro, 1974 | Argentina | Early Cretaceous | Sauropoda: Titanosauridae |
| *Chungkingosaurus* Dong, Zhou & Zhang, 1983 | China | Late Jurassic | Stegosauria: Steosauridae |
| *Cionodon* Cope, 1874 (*nomen dubium*) | | | |
| *Claorhynchus* Cope, 1892 (*nomen dubium*) | | | |
| *Claosaurus* Marsh, 1890 | USA | Late Cretaceous | Ornithopoda: Hadrosauridae |
| *Clasmodosaurus* Ameghino, 1898 (*nomen dubium*) | | | |
| *Clevelanotyrannus* Bakker and others, 1987 (*nomen nudum*) | | | |
| *Coelophysis* Cope, 1889 | USA | Late Triassic | Theropoda: Coelophysidae |
| *Coelosaurus* Leidy, 1865 (*nomen dubium*) | | | |
| *Coeluroides* Huene & Matley, 1933 (*nomen dubium*) | | | |
| *Coelurosaurus* Huene, 1929 (*nomen nudum*) | | | |
| *Coelurus* Marsh, 1879 | USA | Late Triassic | Theropoda: Coelurosauria |
| *Coloradia* Bonaparte, 1978 (*Coloradisaurus*) | | | |
| *Coloradisaurus* Lambert, 1983 | Argentina | Late Triassic | Prosauropoda: Plateosauridae? |
| *Compsognathus* Wagner, 1859 | Germany | Late Jurassic | Theropoda: Compsognathidae |
| *Compsosuchus* Huene & Matley, 1933 (*nomen dubium*) | | | |
| *Conchoraptor* Barsbold, 1986 | Mongolia | Late Cretaceous | Theropoda: Oviraptoridae |
| *Corythosaurus* Brown, 1914 | USA | Late Cretaceous | Ornithopoda: Hadrosauridae |
| *Craspedodon* Dollo, 1883 (*nomen dubium*) | | | |
| *Crataeomus* Seeley, 1881 (*nomen dubium*) | | | |
| *Craterosaurus* Seeley, 1874 (*nomen dubium*) | | | |
| *Creosaurus* Marsh, 1878 (*Allosaurus*) | | | |
| *Cristatusaurus* Taquet & D. A. Russell, 1998 (*nomen dubium*) | | | |
| *Cryolophosaurus* Hammer & Hickerson, 1994 | Antarctica | Early Jurassic | Theropoda: Carnosauria? |
| *Cryptodraco* Lydekker, 1889 (*Cryptosaurus*) | | | |
| *Cryptosaurus* Seeley, 1869 (*nomen dubium*) | | | |
| *Cumnoria* Seeley, 1888 (*Camptosaurus*) | | | |
| *Cylindricodon* Owen/Jaeger, 1828 (*Hylaeosaurus*) | | | |

**D**

| | | | |
|---|---|---|---|
| *Dacentrurus* Lucas, 1902 | England | Late Jurassic | Stegosauria: Stegosauridae |
| *Dachungosaurus* Chao, 1985 (*nomen nudum*) | | | |
| *Damalasaurus* Zhao, 1983 (*nomen nudum*) | | | |
| *Dandakosaurus* Yadagiri, 1982 (*nomen dubium*) | | | |
| *Danubiosaurus* Bunzel, 1871 (*Struthiosaurus*) | | | |

*Daptosaurus* Brown vide Chure & McIntosh, 1989 (*nomen nudum*)

| | | | |
|---|---|---|---|
| *Daspletosaurus* D. A. Russell, 1970 | Canada | Late Cretaceous | Theropoda: Tyrannosauridae |
| *Datousaurus* Dong & Tang, 1984 | China | Mid-Jurassic | Sauropoda: Cetiosauridae? |
| *Deinocheirus* Osmólska & Roniewicz, 1970 | Mongolia | Late Cretaceous | Theropoda: Ornithomimidae |

*Deinodon* Leidy, 1856 (*nomen dubium*)

| | | | |
|---|---|---|---|
| *Deinonychus* Ostrom, 1969 | USA | Early Cretaceous | Theropoda: Dromaeosauridae |
| *Deltadromeus* Sereno and others, 1996 | Morocco | Mid-Cretaceous | Theropoda: Coelurosauria |

*Denversaurus* Bakker, 1988 (*Edmontonia*)
*Dianchungosaurus* Young, 1982 (*nomen dubium*)

| | | | |
|---|---|---|---|
| *Diceratops* Hatcher/ Lull, 1905 | USA | Late Cretaceous | Ceratopsia: Ceratopsidae |

*Diclonius* Cope, 1876 (*nomen dubium*)

| | | | |
|---|---|---|---|
| *Dicraeosaurus* Janensch, 1914 | Tanzania | Late Jurassic | Sauropoda: Diplodocidae |

*Didanodon* Osborn, 1902 (*Trachodon*)

| | | | |
|---|---|---|---|
| *Dilophosaurus* Welles, 1970 | USA | Early Jurassic | Theropoda: Coelophysidae |

*Dimodosaurus* Pidancet & Chopard, 1862 (*Plateosaurus*)

| | | | |
|---|---|---|---|
| *Dinheirosaurus* Bonaparte & Mateus, 1999 | Portugal | Late Jurassic | Sauropoda: Diplodocidae |

*Dinodocus* Owen, 1884 (*Pelorosaurus*)
*Dinosaurus* Rütimeyer, 1856 (*Plateosaurus*)
*Dinotyrannus* Olshevsky, Ford & Yamamoto, 1995 (*Tyrannosaurus*)

| | | | |
|---|---|---|---|
| *Diplodocus* Marsh, 1878 | USA | Late Jurassic | Sauropoda: Diplodocidae |

*Diplotomodon* Leidy, 1868 (*nomen dubium*)
*Diracodon* Marsh, 1881 (*Stegosaurus*)
*Dolichosuchus* Huene, 1932 (*nomen dubium*)
*Doryphorosaurus* Nopcsa, 1916 (*Kentrosaurus*)

| | | | |
|---|---|---|---|
| *Dracopelta* Galton, 1980 | Portugal | Late Jurassic | Ankylosauria: Ankylosauridae |
| *Drinker* Bakker, Galton, Siegwarth & Filla, 1990 | USA | Late Jurassic | Ornithopoda: Hypsilophodontidae |
| *Dromaeosaurus* Matthew & Brown, 1922 | Canada | Late Cretaceous | Theropoda: Dromaeosauridae |
| *Dromiceiomimus* D. A. Russell, 1972 | Canada | Late Cretaceous | Theropoda: Ornithomimidae |

*Dromicosaurus* van Hoepen, 1920 (*Massospondylus*)

| | | | |
|---|---|---|---|
| *Dryosaurus* Marsh, 1894 | USA/Tanzania | Late Jurassic | Ornithopoda: Dryosauridae |

*Dryptosauroides* Huene & Matley, 1933 (*nomen dubium*)

| | | | |
|---|---|---|---|
| *Dryptosaurus* Marsh, 1877 | USA | Late Cretaceous | Theropoda: Coelurosauria |

*Dynamosaurus* Osborn, 1905 (*Tyrannosaurus*)
*Dyoplosaurus* Parks, 1924 (*Euoplocephalus*)
*Dysalotosaurus* Virchow, 1919 (*Dryosaurus*)
*Dysganus* Cope, 1876 (*nomen dubium*)

| | | | |
|---|---|---|---|
| *Dyslocosaurus* McIntosh and others, 1992 | USA | Late Jurassic | Sauropoda: Diplodocidae |

*Dystrophaeus* Cope, 1877 (*nomen dubium*)

| | | | |
|---|---|---|---|
| *Dystylosaurus* Jensen, 1985 | USA | Late Jurassic | Sauropoda: Diplodocidae |

## E

| | | | |
|---|---|---|---|
| *Echinodon* Owen, 1861 | England | Late Jurassic | Ornithischia: Thyreophora? |
| *Edmarka* Bakker, Kralis, Siegwarth & Filla, 1992 | USA | Late Jurassic | Theropoda: Torvosauridae |
| *Edmontonia* C. M. Sternberg, 1928 | Canada | Late Cretaceous | Ankylosauria: Nodosauridae |
| *Edmontosaurus* Lambe, 1917 | Canada | Late Cretaceous | Ornithopoda: Hadrosauridae |

*Efraasia* Galton, 1973 (*Sellosaurus*)

| | | | |
|---|---|---|---|
| *Einiosaurus* Sampson, 1995 | USA | Late Cretaceous | Ceratopsia: Ceratopsidae |
| *Elaphrosaurus* Janensch, 1920 | Tanzania | Late Jurassic | Theropoda: Ceratosauria? |
| *Elmisaurus* Osmólska, 1981 | Mongolia | Late Cretaceous | Theropoda: Elmisauridae |

*Elopteryx* Andrews, 1913 (*nomen dubium*)
*Elosaurus* Peterson & Gilmore, 1902 (*Apatosaurus*)
*Elvisaurus* Holmes, 1993 (*nomen nudum*)

| | | | |
|---|---|---|---|
| *Emausaurus* Haubold, 1991 | Germany | Early Jurassic | Thyreophora: Scelidosauridae |

*Embasaurus* Riabinin, 1931 (*nomen dubium*)

| | | | |
|---|---|---|---|
| *Enigmosaurus* Barsbold & Perle, 1983 | Mongolia | Late Cretaceous | Theropoda: Therizinosauridae |
| *Eobrontosaurus* Bakker, 1998 | USA | Late Jurassic | Sauropoda: Diplodocidae |

*Eoceratops* Lambe, 1915 (*Chasmosaurus*)
*Eohadrosaurus* Kirkland, 1997 (*nomen nudum*)

| | | | |
|---|---|---|---|
| *Eolambia* Kirkland, 1998 | USA | Late Cretaceous | Ornithopoda: Hadrosauridae |
| *Eoraptor* Sereno, Forster, Rogers & Monetta, 1993 | Argentina | Late Triassic | Theropoda: Eoraptoridae? |
| *Epachthosaurus* J. Powell, 1990 | Argentina | Mid-Cretaceous | Sauropoda: Titanosauridae |

*Epanterias* Cope, 1878 (*Allosaurus*)

| | | | |
|---|---|---|---|
| *Erectopus* Huene, 1922 | France | Early Cretaceous | Theropoda: Neotetanurae |
| *Erlikosaurus* Barsbold & Perle, 1980 | Mongolia | Late Cretaceous | Theropoda: Therizinosauridae |

*Euacanthus* Owen & Tennyson, 1897 (*nomen nudum*)
*Eucamerotus* Hulke, 1872 (*nomen dubium*)
*Eucentrosaurus* Chure & McIntosh, 1989 (*Centrosaurus*)

*Eucercosaurus* Seeley, 1879 (*Anoplosaurus*)
*Eucnemesaurus* van Hoepen, 1920 (*Euskelosaurus*)
*Eucoelophysis* Sullivan & Lucas, 1999 (*nomen dubium*)

| | | | |
|---|---|---|---|
| *Euhelopus* Romer, 1956 | China | Late Jurassic | Sauropoda: Eusauropoda |
| *Euoplocephalus* Lambe, 1910 | Canada | Late Cretaceous | Ankylosauria: Ankylosauridae |

*Eureodon* Brown & Olshevsky, 1991 (*nomen nudum*)

| | | | |
|---|---|---|---|
| *Euronychodon* Telles-Antunes & Sigogneau-Russell, 1991 | Portugal | Late Cretaceous | Theropoda: Dromaeosauridae |
| *Euskelosaurus* Huxley, 1866 | South Africa | Late Triassic | Prosauropoda: Plateosauridae? |
| *Eustreptospondylus* Walker, 1964 | England | Late Jurassic | Theropoda: Tetanurae |

**F**

| | | | |
|---|---|---|---|
| *Fabrosaurus* Ginsburg, 1964 | South Africa | Early Jurassic | Ornithopoda: Fabrosauridae |

*Fenestrosaurus* Osborn, 1924 (*nomen nudum*)
*Frenguellisaurus* Novas, 1986 (*Herrerasaurus*)
*Fukuisaurus* Lambert, 1990 (*nomen nudum*)
*Fulengia* Carroll & Galton, 1977 (*Lufengosaurus*)
*Fulgurotherium* Huene, 1932 (*nomen dubium*)
*Futabasaurus* Lambert, 1990 (*nomen nudum*)

**G**

*Gadolosaurus* Saito, 1979 (*nomen nudum*)

| | | | |
|---|---|---|---|
| *Gallimimus* Osmólska, Roniewicz & Barsbold, 1972 | Mongolia | Late Cretaceous | Theropoda: Ornithomimidae |
| *Galtonia* Huber, Lucas & Hunt, 1993 | USA | Late Triassic | Ornithischia: family? |
| *Gargoyleosaurus* Carpenter, Miles & Cloward, 1998 | USA | Late Jurassic | Ankylosauria: Ankylosauridae |
| *Garudimimus* Barsbold, 1981 | Mongolia | Late Cretaceous | Theropoda: Ornithomimidae |
| *Gasosaurus* Dong, 1985 | China | Mid-Jurassic | Theropoda: Coelurosauri |
| *Gasparinisaura* Coria & Salgado, 1996 | Argentina | Late Cretaceous | Ornithopoda: Euiguanodontia |
| *Gastonia* Kirkland, 1998 | USA | Early Cretaceous | Ankylosauria: Ankylosauridae |
| *Genusaurus* Accarie and others, 1995 | France | Early Cretaceous | Theropoda: Neoceratosauria |
| *Genyodectes* Woodward, 1901 | Argentina | Late Cretaceous | Theropoda: Neoceratosauria |

*Geranosaurus* Broom, 1911 (*nomen dubium*)

| | | | |
|---|---|---|---|
| *Giganotosaurus* Coria & Salgado, 1995 | Argentina | Early Cretaceous | Theropoda: Allosauridae |

*Gigantosaurus* Seeley, 1869 (*nomen dubium*)
*Gigantosaurus* E. Fraas, 1908 (*Barosaurus*)
*Gigantoscelus* van Hoepen, 1916 (*Euskelosaurus*)
*Gigantspinosaurus* Anonymous, 1993 (*nomen nudum*)

| | | | |
|---|---|---|---|
| *Gilmoreosaurus* Brett-Surman, 1979 | China | Late Cretaceous | Ornithopoda: Hadrosauridae |
| *Giraffatitan* Paul, 1988 | Tanzania | Late Jurassic | Sauropoda: Brachiosauridae |

*Gobisaurus* Spinar, Currie & Sovak, 1994 (*nomen nudum*)

| | | | |
|---|---|---|---|
| *Gojirasaurus* Carpenter, 1997 | USA | Late Triassic | Theropoda: Coelophysidae |
| *Gondwanatitan* Kellner & de Azevedo, 1999 | Brazil | Late Cretaceous | Sauropoda: Titanosauridae |

*Gongbusaurus* Dong, Zhou & Zhang, 1983 (*nomen dubium*)

| | | | |
|---|---|---|---|
| *Gongxianosaurus* He and others, 1998 | China | Early Jurassic | Prosauropoda: family? |
| *Gorgosaurus* Lambe, 1914 | Canada/USA | Late Cretaceous | Theropoda: Tyrannosauridae |
| *Goyocephale* Perle, Maryanska & Osmólska, 1982 | Mongolia | Late Cretaceous | Pachycephalosauria: Pachycephalosauridae |

*Graciliceratops* Forster & Sereno, 1997 (*nomen nudum*)
*Gravisaurus* Norman, 1989 (*nomen nudum*)

| | | | |
|---|---|---|---|
| *Gravitholus* Wall & Galton, 1979 | Canada | Late Cretaceous | Pachycephalosauria: Pachycephalosauridae |

*Gresslyosaurus* Rütimeyer, 1857 (*Plateosaurus*)
*Gryponyx* Broom, 1911 (*Massospondylus*)

| | | | |
|---|---|---|---|
| *Gryposaurus* Lambe, 1914 | Canada | Late Cretaceous | Ornithopoda: Hadrosauridae |
| *Guaibasaurus* Bonaparte, Ferigolo & Ribeiro, 1999 | Brazil | Late Triassic | Prosauropoda: family? |

*Gyposaurus* Broom, 1911 (*Massospondylus*)

**H**

*Hadrosauravus* Lambert, 1990 (*nomen nudum*)

| | | | |
|---|---|---|---|
| *Hadrosaurus* Leidy, 1859 | USA | Late Cretaceous | Ornithopoda: Hadrosauridae |

*Halticosaurus* Huene, 1908 (*nomen dubium*)

| | | | |
|---|---|---|---|
| *Haplocanthosaurus* Hatcher, 1903 | USA | Late Jurassic | Sauropoda: Cetiosauridae? |

*Haplocanthus* Hatcher, 1903 (*Haplocanthosaurus*)

| | | | |
|---|---|---|---|
| *Harpymimus* Barsbold & Perle, 1984 | Mongolia | Late Cretaceous | Theropoda: Ornithomimidae |

*Hecatasaurus* Brown, 1910 (*Telmatosaurus*)
*Heishansaurus* Bohlin, 1953 (*nomen dubium*)
*Helopus* Wiman, 1929 (*Euhelopus*)
*Heptasteornis* Harrison & C. A. Walker, 1975 (*Elopteryx*)

*Herrerasaurus* Reig, 1963 — Argentina — Late Triassic — Theropoda: Herrerasauridae
*Heterodontosaurus* Crompton & Charig, 1962 — South Africa — Early Jurassic — Ornithopoda: Heterodontosauridae
*Heterosaurus* Cornuel, 1850 (*Iguanodon*)
*Hierosaurus* Wieland, 1909 (*Nodosaurus*)
*Hikanodon* Keferstein, 1834 (*Iguanodon*)
*Hironosaurus* Hisa, 1988 (*nomen nudum*)
*Hisanohamasaurus* Lambert, 1990 (*nomen nudum*)
*Histriasaurus* Dalla Vecchia, 1998 — Croatia — Early Cretaceous — Sauropoda: Diplodocidae
*Homalocephale* Maryanska & Osmólska, 1974 — Mongolia — Late Cretaceous — Pachcephalosauria: Pachycephalosauridae
*Honghesaurus* Anonymous, 1981 (*nomen nudum*)
*Hoplitosaurus* Lucas, 1902 — USA — Late Cretaceous — Ankylosauria: Ankylosauridae
*Hoplosaurus* Seeley, 1881 (*Struthiosaurus*)
*Hortalotarsus* Seeley, 1894 (*nomen dubium*)
*Huabeisaurus* Liu & Pang, 1999 (*nomen nudum*)
*Huayangosaurus* Dong, Tang & Zhou, 1982 — China — Mid-Jurassic — Stegosauria: Huayangosauridae
*Hudiesaurus* Dong, 1997 (*nomen dubium*)
*Hulsanpes* Osmólska, 1982 — Mongolia — Late Cretaceous — Theropoda: Maniraptora
*Hylaeosaurus* Mantell, 1833 — England — Early Cretaceous — Ankylosauridae: Nodosauridae
*Hylosaurus* Fitzinger, 1843 (*Hylaeosaurus*)
*Hypacrosaurus* Brown, 1913 — Canada/USA — Late Cretaceous — Ornithopoda: Hadrosauridae
*Hypselosaurus* Matheron, 1869 — France — Late Cretaceous — Sauropoda: Titanosauridae
*Hypsibema* Cope, 1869 (*nomen dubium*)
*Hypsilophodon* Huxley, 1869 — England — Early Cretaceous — Ornithopoda: Hypsilophodontidae
*Hypsirophus* Cope, 1878 (*Allosaurus*)

# I

*Iguanodon* Mantell, 1825 — England — Early Cretaceous — Ornithopoda: Iguanodontidae
*Iguanosaurus* Anonymous, 1824 (*nomen nudum*)
*Iliosuchus* Huene, 1932 — England — Mid-Jurassic — Theropoda: Tyrannosauroidea
*Ilokelesia* Coria & Salgado, 1999 (*nomen nudum*)
*Indosaurus* Huene & Matley, 1933 — India — Late Cretaceous — Theropoda: Abelisauridae
*Indosuchus* Huene & Matley, 1933 — India — Late Cretaceous — Theropoda: Abelisauridae
*Ingenia* Barsbold, 1981 — Mongolia — Late Cretaceous — Theropoda: Oviraptoridae
*Inosaurus* de Lapparent, 1960 (*nomen dubium*)
*Irritator* Martill and others, 1996 — Brazil — Mid-Cretaceous — Theropoda: Spinosauridae
*Ischisaurus* Reig, 1963 (*Herrerasaurus*)
*Ischyrosaurus* Hulke, 1874 (*nomen dubium*)
*Itemirus* Kurzanov, 1976 — Uzbekistan — Late Cretaceous — Theropoda: Tyrannosauridae
*Iuticosaurus* Le Loeuff, 1993 (*nomen dubium*)

# J

*Jainosaurus* Hunt and others, 1995 — India — Late Cretaceous — Sauropoda: Titanosauridae
*Janenschia* Wild, 1991 — Tanzania — Late Jurassic — Sauropoda: Titanosauridae
*Jaxartosaurus* Riabinin, 1937 — Kazakhstan — Late Cretaceous — Ornithopoda: Hadrosauridae
*Jenghizkhan* Olshevsky, 1995 (*Tarbosaurus*)
*Jensenosaurus* Olshevskye, 1996 (*nomen nudum*)
*Jiangjunmiaosaurus* Anonymous, 1987 (*nomen nudum*)
*Jingshanosaurus* Zhang & Yang, 1995 — China — Early Jurassic — Prosauropoda: Yunnanosauridae?
*Jobaria* Sereno and others, 1999 — Niger — Mid-Cretaceous — Sauropoda: Cetiosauridae?
*Jubbulpuria* Huene & Matley, 1933 (*nomen dubium*)
*Jurassosaurus* Dong, 1992 (*nomen nudum*)

# K

*Kagasaurus* Hisa, 1988 (*nomen nudum*)
*Kaijiangosaurus* He, 1984 — China — Middle Jurassic — Theropoda: Tetanurae
*Kakuru* Molnar & Pledge, 1980 — Australia — Early Cretaceous — Theropoda: Avimimidae?
*Kangnasaurus* Haughton, 1915 (*nomen dubium*)
*Katsuyamasaurus* Lambert, 1990 (*nomen nudum*)
*Kelmayisaurus* Dong, 1973 (*nomen dubium*)
*Kentrosaurus* Hennig, 1915 — Tanzania — Late Jurassic — Stegosauria: Stegosauridae
*Kentrurosaurus* Hennig, 1916 (*Kentrosaurus*)
*Kitadanisaurus* Lambert, 1990 (*nomen nudum*)
*Klamelisaurus* Zhao, 1993 — China — Late Jurassic — Sauropoda: Eusauropoda
*Koparion* Chure, 1994 — USA — Late Jurassic — Theropoda: Maniraptoriformes
*Koreanosaurus* Kim, 1979 (*nomen nudum*)
*Kotasaurus* Yadagiri, 1988 — India — Early Jurassic — Sauropoda: Vulcanodontidae?

*Kritosaurus* Brown, 1910 — Canada/USA — Late Cretaceous — Ornithopoda: Hadrosauridae
*Kulceratops* Nesov, 1995 (*nomen dubium*)
*Kunmingosaurus* Chao, 1985 (*nomen nudum*)


**L**
*Labocania* Molnar, 1974 (*nomen dubium*)
*Labrosaurus* Marsh, 1879 (*Allosaurus*)
*Laelaps* Cope, 1866 (*Dryptosaurus*)
*Laevisuchus* Huene & Matley, 1933 — India — Late Cretaceous — Theropoda: Abelisauridae?
*Lambeosaurus* Parks, 1923 — Canada/USA — Late Cretaceous — Ornithopoda: Hadrosauridae
*Lametasaurus* Matley, 1923 (*Indosuchus*)
*Lanasaurus* Gow, 1975 — South Africa — Early Jurassic — Ornithopoda: Heterodontisauridae
*Lancangosaurus* Dong, Zhou & Zhang, 1983 (*nomen nudum*)
*Lancanjiangosaurus* Zhao, 1983 (*nomen nudum*)
*Laosaurus* Marsh, 1878 (*Othnielia*)
*Laplatasaurus* Huene, 1929 — Argentina — Late Cretaceous — Sauropoda: Titanosauridae
*Lapparentosaurus* Bonaparte, 1986 — Madagascar — Mid-Jurassic — Sauropoda: Eusauropoda
*Leaellynasaura* T. Rich & P. Rich, 1989 — Australia — Early Cretaceous — Ornithopoda: Hypsilophodontidae
*Leipsanosaurus* Nopcsa, 1918 (*Struthiosaurus*)
*Leptoceratops* Brown, 1914 — Canada/USA — Late Cretaceous — Ceratopsia: Ceratopsidae
*Leptospondylus* Owen, 1854 (*Massospondylus*)
*Lesothosaurus* Galton, 1978 — South Africa — Early Jurassic — Ornithopoda: Fabrosauridae
*Lessemsaurus* Bonaparte, 1999 — Argentina — Late Triassic — Prosauropoda: Melanorosauridae?
*Lexovisaurus* Hoffstetter, 1957 — France — Mid-Jurassic — Stegosauria: Stegosauridae
*Liassaurus* Welles, H. P. Powell & Pickering, 1995 (*nomen nudum*)
*Liqabueino* Bonaparte, 1996 — Argentina — Early Cretaceous — Theropoda: Abelisauridae?
*Liliensternus* Welles, 1984 — Germany — Late Triassic — Theropoda: Coelophysidae
*Limaysaurus* Calvo & Salgado, 1997 (*nomen nudum*)
*Limnosaurus* Nopcsa, 1899 (*Telmatosaurus*)
*Lirainosaurus* Sanz and others, 1999 — Spain — Late Cretaceous — Sauropoda: Titanosauridae
*Loncosaurus* Ameghino, 1898 (*nomen dubium*)
*Longosaurus* Welles, 1984 (*Coelophysis*)
*Lophorhothon* Langston, 1960 — USA — Late Cretaceous — Ornithopoda: Hadrosauridae
*Loricosaurus* Huene, 1929 (*Neuquensaurus*)
*Lourinhanosaurus* Mateus, 1998 — Portugal — Late Jurassic — Theropoda: Allosauroidea
*Lourinhasaurus* Dantas and others, 1998 — Portugal — Late Jurassic — Sauropoda: Macronaria
*Luanpingosaurus* Cheng, 1996 (*nomen nudum*)
*Lucianosaurus* Hunt & Lucas, 1994 — USA — Late Triassic — Ornithischia: family?
*Lufengocephalus* Young, 1974 (*Tawasaurus*)
*Lufengosaurus* Young, 1941 — China — Early Jurassic — Prosauropoda: Plateosauridae
*Lurdusaurus* Taquet & D. A. Russell, 1999 — Niger — Mid-Cretaceous — Ornithopoda: Iguanodontidae
*Lusitanosaurus* de Lapparent & Zbyszewski, 1957 (*nomen dubium*)
*Lycorhinus* Haughton, 1924 — South Africa — Early Jurassic — Ornithopoda: Heterodontosauridae


**M**
*Macrodontophion* Zborzewski, 1834 (*nomen dubium*)
*Macrophalangia* C. M. Sternberg, 1932 (*Chirostenotes*)
*Macrurosaurus* Seeley, 1876 — England — Early Cretaceous — Sauropoda: Titanosauria
*Madsenius* Lambert, 1990 (*nomen nudum*)
*Magnosaurus* Huene, 1932 — England — Mid-Jurassic — Theropoda: Tetanurae
*Magulodon* Kranz, 1996 (*nomen nudum*)
*Magyarosaurus* Huene, 1932 — Hungary — Late Cretaceous — Sauropoda: Titanosauridae
*Maiasaura* Horner & Makela, 1979 — USA — Late Cretaceous — Ornithopoda: Hadrosauridae
*Majungasaurus* Lavocat, 1955 (*nomen dubium*)
*Majungatholus* Sues & Taquet, 1979 — Madagascar — Late Cretaceous — Theropoda: Abelisauridae
*Malawisaurus* Jacobs and others, 1993 — Malawi — Early Cretaceous — Sauropoda: Titanosauridae
*Maleevosaurus* Carpenter, 1992 (*Tarbosaurus*)
*Maleevus* Tumanova, 1987 — Mongolia — Late Cretaceous — Ankylosauria: Ankylosauridae
*Mamenchisaurus* Young, 1954 — China — Late Jurassic — Sauropoda: Eusaruopoda
*Mandschurosaurus* Riabinin, 1930 — China — Late Cretaceous — Ornithopoda: Iguanodontidae
*Manospondylus* Cope, 1892 (*Tyrannosaurus*)
*Marmarospondylus* Owen, 1875 (*Bothriospondylus*)
*Marshosaurus* Madsen, 1976 — USA — Late Jurassic — Theropoda: Neotetanurae
*Massospondylus* Owen, 1854 — South Africa — Late Triassic — Prosauropoda: Massospondylidae
*Megacervixosaurus* Zhao, 1983 (*nomen nudum*)
*Megadactylus* Hitchcock, 1865 (*Anchisaurus*)
*Megadontosaurus* Brown in Ostrom, 1970 (*nomen nudum*)

| | | | |
|---|---|---|---|
| *Megalosaurus* Buckland, 1824 | England | Mid-Jurassic | Theropoda: Megalosauridae |
| *Megaraptor* Novas, 1998 | Argentina | Late Cretaceous | Theropoda: Maniraptora |
| *Melanorosaurus* Haughton, 1924 | South Africa | Late Triassic | Prosauropoda: Melanorosauridae |
| *Merosaurus* Welles, H. P. Powell & Pickering, 1995 (*nomen nudum*) | | | |
| *Metriacanthosaurus* Walker, 1964 | England | Late Jurassic | Theropoda: Tetanurae |
| *Microcephale* Sereno, 1997 (*nomen nudum*) | | | |
| *Microceratops* Bohlin, 1953 | China | Late Cretaceous | Ceratopsia: Ceratopsidae |
| *Microcoelus* Lydekker, 1893 (*Neuquensaurus*) | | | |
| *Microdontosaurus* Zhao, 1983 (*nomen nudum*) | | | |
| *Microhadrosaurus* Dong, 1979 (*nomen dubium*) | | | |
| *Micropachycephalosaurus* Dong, 1978 | China | Late Cretaceous | Pachycephalosauria: Pachycephalosauridae |
| *Microsaurops* Kuhn, 1963 (*Neuquensaurus*) | | | |
| *Microvenator* Ostrom, 1970 | USA | Early Cretaceous | Theropoda: Maniraptor |
| *Mifunesaurus* Hisa, 1985 (*nomen nudum*) | | | |
| *Minmi* Molnar, 1980 | Australia | Early Cretaceous | Ankylosauria: Nodosauridae |
| *Mochlodon* Seeley, 1881 (*Rhabdodon*) | | | |
| *Mongolosaurus* Gilmore, 1933 (*nomen dubium*) | | | |
| *Monkonosaurus* Zhao, 1990 | China | Late Jurassic | Stegosauria: Stegosauridae |
| *Monoclonius* Cope, 1876 (*nomen dubium*) | | | |
| *Monolophosaurus* Zhao & Currie, 1994 | China | Late Jurassic | Theropoda: Carnosauria |
| *Mononychus* Perle, Norell, Chiappe & Clark, 1993 (*Mononykus*) | | | |
| *Mononykus* Perle, Norell, Chiappe & Clark, 1993 | Mongolia | Late Cretaceous | Theropoda: Alvarezsauria |
| *Montanoceratops* C. M. Sternberg, 1951 | USA | Late Cretaceous | Ceratopsia: Ceratopsidae |
| *Morinosaurus* Sauvage, 1874 (*nomen dubium*) | | | |
| *Morosaurus* Marsh, 1878 (*Camarasaurus*) | | | |
| *Moshisaurus* Hisa, 1985 (*nomen nudum*) | | | |
| *Mussaurus* Bonaparte & Vince, 1979 | Argentina | Late Triassic | Prosauropoda: family? |
| *Muttaburrasaurus* Bartholomai & Molnar, 1981 | Australia | Early Cretaceous | Ornithopoda: Iguanodontia |
| *Mymoorapelta* Kirkland & Carpenter, 1994 | USA | Late Jurassic | Ankylosauria: Ankylosauridae |

### N

| | | | |
|---|---|---|---|
| *Naashoibitosaurus* Hunt & Lucas, 1993 | USA | Late Cretaceous | Ornithopoda: Hadrosauridae |
| *Nanosaurus* Marsh, 1877 (*nomen dubium*) | | | |
| *Nanotyrannus* Bakker, Williams & Currie, 1988 (*Tyrannosaurus*) | | | |
| *Nanshiungosaurus* Dong, 1979 | China | Late Cretaceous | Theropoda: Therizinosauridae |
| *Nectosaurus* Versluys, 1910 (*Kritosaurus*) | | | |
| *Nedcolbertia* Kirkland and others, 1998 | USA | Early Cretaceous | Theropoda: Maniraptoriformes |
| *Nemegtosaurus* Nowinski, 1971 | Mongolia | Late Cretaceous | Sauropoda: Nemegtosauridae |
| *Neosaurus* Gilmore, 1945 (*nomen dubium*) | | | |
| *Neosodon* de la Moussaye, 1885 (*Pelorosaurus*) | | | |
| *Neovenator* Hutt, Martill & Barker, 1996 | England | Early Cretaceous | Theropoda: Allosauridae |
| *Neuquensaurus* J. Powell, 1992 | Argentina | Late Cretaceous | Sauropoda: Saltasauridae |
| *Newtonsaurus* Welles,, 1999 (*nomen nudum*) | | | |
| *Ngexisaurus* Zhao, 1983 (*nomen nudum*) | | | |
| *Nigersaurus* Sereno and others, 1999 | Niger | Mid-Cretaceous | Sauropoda: Rebbachisauridae |
| *Niobrarasaurus* Carpenter and others, 1995 | USA | Late Cretaceous | Ankylosauria: Nodosauridae |
| *Nipponosaurus* Nagao, 1936 | Russia | Late Cretaceous | Ornithopoda: Hadrosauridae |
| *Noasaurus* Bonaparte & J. Powell, 1980 | Argentina | Late Cretaceous | Theropoda: Abelisauria |
| *Nodocephalosaurus* Sullivan, 1999 | USA | Late Cretaceous | Ankylosauria: Ankylosauridae |
| *Nodosaurus* Marsh, 1889 | USA | Late Cretaceous | Ankylosauria: Nodosauridae |
| *Nomingia* Barsbold and others, 2000 | Mongolia | Late Cretaceous | Theropda: Oviraptoridae |
| *Notoceratops* Tapia, 1918 (*nomen dubium*) | | | |
| *Notohypsilophodon* Martínez, 1998 | Argentina | Late Cretaceous | Ornithopoda: family? |
| *Nurosaurus* Dong, 1992 (*nomen nudum*) | | | |
| *Nuthetes* Owen, 1854 (*nomen dubium*) | | | |

### O

| | | | |
|---|---|---|---|
| *Ohmdenosaurus* Wild, 1978 | Germany | Early Jurassic | Sauropoda: Vulcanodontidae |
| *Oligosaurus* Seeley, 1881 (*Rhabdodon*) | | | |
| *Omeisaurus* Young, 1939 | China | Late Jurassic | Sauropoda: Eusauropoda |
| *Omosaurus* Owen, 1875 (*Dacentrurus*) | | | |
| *Onychosaurus* Nopcsa, 1902 (*Struthiosaurus*) | | | |
| *Opisthocoelicaudia* Borsuk-Bialynicka, 1977 | Mongolia | Late Cretaceous | Sauropoda: Saltasauridae |
| *Oplosaurus* Gervais, 1852 (*Pelorosaurus*) | | | |
| *Orcomimus* Triebold, 1997 (*nomen nudum*) | | | |
| *Orinosaurus* Lydekker, 1889 (*Euskelosaurus*) | | | |

| | | | |
|---|---|---|---|
| *Ornatotholus* Galton & Sues, 1983 | Canada/USA | Late Cretaceous | Pachycephalosauria: Pachycephalosauridae |
| *Ornithodesmus* Seeley, 1887 (*nomen dubium*) | | | |
| *Ornithoides* Osborn, 1924 (*nomen nudum*) | | | |
| *Ornitholestes* Osborn, 1903 | USA | Late Jurassic | Theropoda: Maniraptoriformes |
| *Ornithomerus* Seeley, 1881 (*Rhabdodon*) | | | |
| *Ornithomimoides* Huene & Matley, 1933 (*nomen dubium*) | | | |
| *Ornithomimus* Marsh, 1890 | USA | Late Cretaceous | Theropoda: Ornithomimidae |
| *Ornithopsis* Seeley, 1870 (*nomen dubium*) | | | |
| *Ornithotarsus* Cope, 1869 (*nomen dubium*) | | | |
| *Orodromeus* Horner & Weishampel, 1988 | USA | Late Cretaceous | Ornithopoda: Hypsilophodontidae |
| *Orosaurus* Huxley, 1867 (*Euskelosaurus*) | | | |
| *Orthogoniosaurus* Das-Gupta, 1931 (*Indosaurus*) | | | |
| *Orthomerus* Seeley, 1883 (*nomen dubium*) | | | |
| *Oshanosaurus* Chao, 1985 (*nomen nudum*) | | | |
| *Othnielia* Galton, 1977 | USA | Late Jurassic | Ornithopoda: Hypsilophodontidae? |
| *Ouranosaurus* Taquet, 1976 | Niger | Mid-Cretaceous | Ornithopoda: Iguanodontidae? |
| *Oviraptor* Osborn, 1924 | Mongolia | Late Cretaceous | Theropoda: Oviraptoridae |
| *Ovoraptor* Osborn, 1924 (*nomen nudum*) | | | |
| *Ozraptor* Long & Molnar, 1998 | Australia | Mid-Jurassic | Theropoda: Neotheropoda |

**P**

| | | | |
|---|---|---|---|
| *Pachycephalosaurus* Brown & Schlaikjer, 1943 | USA/Canada | Late Cretaceous | Pachycephalosauria: Pachycephalosauridae |
| *Pachyrhinosaurus* C. M. Sternberg, 1950 | Canada | Late Cretaceous | Ceratopsia: Ceratopsidae |
| *Pachysauriscus* Kuhn, 1959 (*Plateosaurus*) | | | |
| *Pachysaurops* Huene, 1961 (*Plateosaurus*) | | | |
| *Pachysaurus* Huene, 1908 (*Pachysauriscus*) | | | |
| *Pachyspondylus* Owen, 1854 (*Massospondylus*) | | | |
| *Palaeopteryx* Jensen, 1981 (*nomen dubium*) | | | |
| *Palaeoscincus* Leidy, 1856 (*nomen dubium*) | | | |
| *Panoplosaurus* Lambe, 1919 | Canada/USA | Late Cretaceous | Ankylosauria: Nodosauridae |
| *Paraiguanodon* Brown in Olshevsky, 1991 (*nomen nudum*) | | | |
| *Paranthodon* Nopcsa, 1929 | South Africa | Jur./Cret. bound | Stegosauria: Stegosauridae |
| *Pararhabdodon* Casanovas-Cladellas and others, 1993 | Spain | Late Cretaceous | Ornithopoda: Iguanodontia |
| *Parasaurolophus* Parks, 1922 | Canada/USA | Late Cretaceous | Ornithopoda: Hadrosauridae |
| *Parksosaurus* C. M. Sternberg, 1937 | Canada | Late Cretaceous | Ornithopoda: Hypsilophodontidae |
| *Paronychodon* Cope, 1876 (*nomen dubium*) | | | |
| *Parrosaurus* Gilmore, 1945 (*nomen dubium*) | | | |
| *Parvicursor* Karhu & Rautian, 1996 | Mongolia | Late Cretaceous | Theropoda: Alvarezsauridae |
| *Patagonykus* Novas, 1996 | Argentina | Late Cretaceous | Theropoda: Alvarezsauridae |
| *Patagosaurus* Bonaparte, 1979 | Argentina | Mid-Jurassic | Sauropoda: Eusauropoda |
| *Patricosaurus* Seeley, 1887 (*nomen dubium*) | | | |
| *Pawpawsaurus* Lee, 1996 | USA | Mid-Cretaceous | Ankylosauria: Nodosauridae |
| *Pectinodon* Carpenter, 1982 (*Troodon*) | | | |
| *Peishansaurus* Bohlin, 1953 (*nomen dubium*) | | | |
| *Pekinosaurus* Hunt & Lucas, 1994 | USA | Late Triassic | Ornithischia: family? |
| *Pelecanimimus* Perez-Moreno and others, 1994 | Spain | Early Cretaceous | Theropoda: Ornithomimidae |
| *Pellegrinisaurus* Salgado, 1996 | Argentina | Late Cretaceous | Sauropoda: Titanosauria |
| *Pelorosaurus* Mantell, 1850 | England | Early Cretaceous | Sauropoda: Titanosauriformes |
| *Peltosaurus* Brown in Chure & McIntosh, 1989 (*Sauropelta*) | | | |
| *Pentaceratops* Osborn, 1923 | USA | Late Cretaceous | Ceratopsia: Ceratopsidae |
| *Phaedrolosaurus* Dong, 1973 (*nomen dubium*) | | | |
| *Phuwiangosaurus* Martin and others, 1994 | Thailand | Early Cretaceous | Sauropoda: Nemegtosauridae |
| *Phyllodon* Thulborn, 1973 (*nomen dubium*) | | | |
| *Piatnitzkysaurus* Bonaparte, 1979 | Argentina | Late Jurassic | Theropoda: Tetanurae |
| *Picrodon* Seeley, 1898 (*nomen dubium*) | | | |
| *Pinacosaurus* Gilmore, 1933 | Mongolia | Late Cretaceous | Ankylosauria: Ankylosauridae |
| *Pisanosaurus* Casamiquela, 1967 | Argentina | Late Triassic | Ornithischia: family? |
| *Piveteausaurus* Taquet & Welles, 1977 | France | Late Jurassic | Theropoda: Coelurosauria |
| *Plateosauravus* Huene, 1932 (*Melanorosaurus*) | | | |
| *Plateosaurus* von Meyer, 1837 | Germany | Late Triassic | Prosauropoda: Plateosauridae |
| *Pleurocoelus* Marsh, 1888 | USA | Early Cretaceous | Sauropoda: Titanosauriformes |
| *Pleuropeltis* Seeley, 1881 (*nomen dubium*) | | | |
| *Podokesaurus* Talbot, 1911 | USA | Late Triassic | Theropoda: Coelophysidae |
| *Poekilopleuron* Eudes-Deslongchamps, 1838 | France | Mid-Jurassic | Theropoda: Torvosauridae |

*Polacanthoides* Nopcsa, 1928 (*Polacanthus*)
*Polacanthus* Owen, 1865      England     Early Cretaceous    Ankylosauria: Nodosauridae
*Polyodontosaurus* Gilmore, 1932 (*Troodon*)
*Polyonax* Cope, 1874 (*nomen dubium*)
*Ponerosteus* Olshevsky, 2000 (*nomen dubium*)
*Prenocephale* Maryanska & Osmólska, 1974    Mongolia    Late Cretaceous    Pachycephalosauria:
                                          Pachycephalosauridae

*Priconodon* Marsh, 1888 (*nomen dubium*)
*Priodontognathus* Seeley, 1875 (*nomen dubium*)
*Probactrosaurus* Rozhdestvenskii, 1966     China      Early Cretaceous    Ornithopoda: Iguanodontia
*Proceratops* Lull, 1906 (*Ceratops*)
*Proceratosaurus* Huene, 1926       England     Mid-Jurassic     Theropoda: Coelurosauria
*Procerosaurus* Fritsch, 1905 (*Ponerosteus*)
*Procheneosaurus* Matthew, 1920 (*Lambeosaurus*)
*Procompsognathus* E. Fraas, 1913     Germany    Late Triassic     Theropoda: Coelophysidae
*Prodeinodon* Osborn, 1924 (*nomen dubium*)
*Proiguanodon* van den Broeck, 1900 (*nomen nudum*)
*Prosaurolophus* Brown, 1916       Canada/USA   Late Cretaceous    Ornithopoda: Hadrosauridae
*Protarchaeopteryx* Ji Q. & Ji S., 1997     China      Early Cretaceous    Theropoda: Maniraptora
*Protiguanodon* Osborn, 1923 (*Psittacosaurus*)
*Protoavis* Chatterjee, 1991        USA       Late Triassic     Theropoda: family?
*Protoceratops* Granger & Gregory, 1923    Mongolia    Late Cretaceous    Ceratopsia: Protoceratopsidae
*Protognathosaurus* Olshevsky, 1991     China      Early Cretaceous    Sauropoda: Eusauropoda
*Protognathus* Zhang, 1988 (*Protognathosaurus*)
*Protohadros* Head, 1998         USA       Mid-Cretaceous    Ornithopoda: Hadrosauridae
*Protorosaurus* Lambe, 1914 (*Chasmosaurus*)
*Protrachodon* Nopcsa, 1923 (*nomen nudum*)
*Psittacosaurus* Osborn, 1923       Mongolia    Early Cretaceous    Ceratopsia: Psittacosauridae
*Pteropelyx* Cope, 1889 (*nomen dubium*)
*Pterospondylus* (*nomen dubium*)

## Q

*Qantassaurus* Rich & Vickers-Rich, 1999    Australia    Early Cretaceous    Ornithopoda: family?
*Qinlingosaurus* Xue, Zhang & Bi, 1996    China      Late Cretaceous    Sauropoda: family?
*Quaesitosaurus* Bannikov & Kurzanov, 1983   Mongolia    Late Cretaceous    Sauropoda: Nemegtosauridae

## R

*Rahona* Forster and others, 1998 (*Rahonavis*)
*Rahonavis* Forster and others, 1998     Madagascar   Late Cretaceous    Theropoda: Avialae
*Rapator* Huene, 1932 (*nomen dubium*)
*Rayososaurus* Bonaparte, 1996 (*nomen dubium*)
*Rebbachisaurus* Lavocat, 1954       Morocco    Mid-Cretaceous    Sauropoda: Rebbachisauridae
*Regnosaurus* Mantell, 1848 (*nomen dubium*)
*Revueltosaurus* Hunt, 1989 (*nomen dubium*)
*Rhabdodon* Matheron, 1869        France      Late Cretaceous    Ornithopoda: Iguanodontia
*Rhadinosaurus* Seeley, 1881 (*nomen dubium*)
*Rhodanosaurus* Nopcsa, 1929 (*Struthiosaurus*)
*Rhoetosaurus* Longman, 1925       Australia    Mid-Jurassic     Sauropoda: Eusauropoda
*Ricardoestesia* Currie, Rigby & Sloan, 1990   Canada/USA   Late Cretaceous    Theropoda: Maniraptoriformes
*Rinchenia* Barsbold, 1997 (*nomen nudum*)
*Rioarribasaurus* Hunt & Lucas, 1991 (*Coelophysis*)
*Riojasaurus* Bonaparte, 1969       Argentina    Late Triassic     Prosauropoda: Melanorosauridae
*Roccosaurus* van Heerden, 1978 (*nomen nudum*)

## S

*Saichania* Maryanska, 1977        Mongolia    Late Cretaceous    Ankylosauria: Ankylosauridae
*Saltasaurus* Bonaparte & J. E. Powell, 1980   Argentina    Late Cretaceous    Sauropoda: Saltasauridae
*Saltopus* Huene, 1910         Scotland    Late Triassic     Theropoda: family?
*Sanchusaurus* Hisa, 1985 (*nomen nudum*)
*Sangonghesaurus* Zhao, 1983 (*nomen nudum*)
*Sanpasaurus* Young, 1944 (*nomen dubium*)
*Santanaraptor* Kellner, 1999       Brazil      Mid-Cretaceous    Theropoda: Maniraporiformes
*Sarcolestes* Lydekker, 1893       England     Mid-Jurassic     Ankylosauria: Nodosauridae
*Sarcosaurus* Andrews, 1921       England     Early Jurassic    Theropoda: Neoceratosauria
*Saturnalia* Langer, Abdala, Richter & Benton, 1999   Brazil      Late Triassic     Prosauropoda: family?
*Sauraechinodon* Falconer, 1861 (*nomen nudum*)
*Saurolophus* Brown, 1912        Canada     Late Cretaceous    Ornithopoda: Hadrosauridae

# FASCINATING FACTS

| | | | |
|---|---|---|---|
| *Sauropelta* Ostrom, 1970 | USA | Early Cretaceous | Ankylosauria: Nodosauridae |
| *Saurophaganax* Chure, 1995 | USA | Late Jurassic | Theropoda: Allosauridae |
| *Saurophagus* Stovall, 1941 (*Allosaurus*) | | | |
| *Sauroplites* Bohlin, 1953 (*nomen dubium*) | | | |
| *Sauroposeidon* Wedel, Cifelli & Sanders, 2000 | USA | Early Cretaceous | Sauropoda: Titanosauriformes |
| *Saurornithoides* Osborn, 1924 | Mongolia | Late Cretaceous | Theropoda: Troodontidae |
| *Saurornitholestes* Sues, 1978 | Canada | Late Cretaceous | Theropoda: Dromaeosauridae |
| *Scelidosaurus* Owen, 1859 | England | Early Jurassic | Thyreophora: Scelidosauridae |
| *Scipionyx* Dal Sasso & Signore, 1998 | Italy | Early Cretaceous | Theropoda: Maniraptoriformes |
| *Scolosaurus* Nopcsa, 1928 (*Euoplocephalus*) | | | |
| *Scrotum* Brookes, 1763 (*Megalosaurus*) | | | |
| *Scutellosaurus* Colbert, 1981 | USA | Early Jurassic | Thyreophora: Scelidosauridae |
| *Secernosaurus* Brett-Surman, 1979 | Argentina | Late Cretaceous | Ornithopoda: Hadrosauridae |
| *Segisaurus* Camp, 1936 | USA | Early Jurassic | Theropoda: Coelophysidae |
| *Segnosaurus* Perle, 1979 | Mongolia | Late Cretaceous | Theropoda: Therizinosauridae |
| *Seismosaurus* Gillette, 1991 | USA | Late Jurassic | Sauropoda: Diplodocidae |
| *Sellosaurus* Huene, 1908 | Germany | Late Triassic | Prosauropoda: Plateosauridae? |
| *Shamosaurus* Tumanova, 1983 | Mongolia | Late Cretaceous | Ankylosauria: Ankylosauridae |
| *Shanshanosaurus* Dong, 1977 | China | Late Cretaceous | Theropoda: Tyrannosauridae |
| *Shantungosaurus* Hu, 1973 | China | Late Cretaceous | Ornithopoda: Hadrosauridae |
| *Shanxia* Barrett, You, Upchurch & Burton, 1998 | China | Late Cretaceous | Ankylosauria: Ankylosauridae |
| *Shanyangosaurus* Xue, Zhang & Bi, 1996 | China | Late Cretaceous | Theropoda: family? |
| *Shunosaurus* Dong, Zhou & Zhang, 1983 | China | Mid-Jurassic | Sauropoda: family? |
| *Shuvuuia* Chiappe, Norell & Clark, 1998 | Mongolia | Late Cretaceous | Theropoda: Alvarezsauridae |
| *Siamosaurus* Buffetaut & Ingavat, 1986 | Thailand | Early Cretaceous | Theropoda: Spinosauria |
| *Siamotyrannus* Buffetaut, Suteethorn & Tong, 1996 | Thailand | Early Cretaceous | Theropoda: Tyrannosauridae |
| *Sigilmassasaurus* D. A. Russell, 1996 (*nomen dubium*) | | | |
| *Siluosaurus* Dong, 1997 | China | Early Cretaceous | Ornithopoda: family? |
| *Silvisaurus* Eaton, 1960 | USA | Early Cretaceous | Ankylosauria: Nodosauridae |
| *Sinocoelurus* Young, 1942 (*nomen dubium*) | | | |
| *Sinornithoides* D. A. Russell & Dong, 1994 | China | Early Cretaceous | Theropoda: Troodontidae |
| *Sinornithosaurus* Xu, Wang & Wu, 1999 | China | Early Cretaceous | Theropoda: Deinonychosauria |
| *Sinosauropteryx* Ji Q. & Ji S., 1996 | China | Early Cretaceous | Theropoda: Compsognathidae |
| *Sinosaurus* Young, 1948 (*nomen dubium*) | | | |
| *Sinraptor* Currie & Zhao, 1994 | China | Late Jurassic | Theropoda: Sinraptoridae |
| *Smilodon* Plieninger, 1846 (*Plateosaurus*) | | | |
| *Sonorasaurus* Ratkevich, 1998 | USA | Early Cretaceous | Sauropoda: Titanosauriformes |
| *Sphenospondylus* Seeley, 1882 (*Iguanodon*) | | | |
| *Spinosaurus* Stromer, 1915 | Egypt | Mid-Cretaceous | Theropoda: Spinosauridae |
| *Spondylosoma* Huene, 1942 (*nomen dubium*) | | | |
| *Staurikosaurus* Colbert, 1970 | Argentina | Late Triassic | Theropoda: Herrerasauridae |
| *Stegoceras* Lambe, 1902 | Canada/USA | Late Cretaceous | Pachycephalosauria: Pachycephalosauridae |
| *Stegopelta* Williston, 1905 (*Nodosaurus*) | | | |
| *Stegosaurides* Bohlin, 1953 (*nomen dubium*) | | | |
| *Stegosaurus* Marsh, 1877 | USA | Late Jurassic | Stegosauria: Stegosauridae |
| *Stenonychosaurus* C. M. Sternberg, 1932 (*Troodon*) | | | |
| *Stenopelix* von Meyer, 1857 | Germany | Early Cretaceous | Ornithischia: family? |
| *Stenotholus* Giffin, Gabriel & Johnson, 1988 (*Stygimoloch*) | | | |
| *Stephanosaurus* Lambe, 1914 (*Kritosaurus*) | | | |
| *Stereocephalus* Lambe, 1902 (*Euoplocephalus*) | | | |
| *Sterrholophus* Marsh, 1891 (*Triceratops*) | | | |
| *Stokesosaurus* Madsen, 1974 | USA | Late Jurassic | Theropoda: Tyrannosauroidea |
| *Strenusaurus* Bonaparte, 1969 (*Riojasaurus*) | | | |
| *Streptospondylus* von Meyer, 1830 | England | Mid-Jurassic | Theropoda: Tetanurae |
| *Struthiomimus* Osborn, 1916 | Mongolia | Late Cretaceous | Theropoda: Ornithomimidae |
| *Struthiosaurus* Bunzel, 1870 | Austria | Late Cretaceous | Ankylosauria: Nodosauridae |
| *Stygimoloch* Galton & Sues, 1983 | USA | Late Cretaceous | Pachycephalosauria: Pachycephalosauridae |
| *Stygivenator* Olshevsky, 1995 | USA | Late Cretaceous | Theropoda: Tyrannosauridae |
| *Styracosaurus* Lambe, 1913 | Canada/USA | Late Cretaceous | Ceratopsia: Ceratopsidae |
| *Suchomimus* Sereno and others, 1998 | Niger | Mid-Cretaceous | Theropoda: Spinosauridae |
| *Sugiyamasaurus* Lambert, 1990 (*nomen nudum*) | | | |
| *Supersaurus* Jensen, 1985 | USA | Late Jurassic | Sauropoda: Diplodocidae |
| *Symphyrophus* Cope, 1878 (*Brachyrophus*) | | | |
| *Syngonosaurus* Seeley, 1879 (*Acanthopholis*) | | | |
| *Syntarsus* Raath, 1969 | Zimbabwe | Early Jurassic | Theropoda: Coelophysidae |

*Syrmosaurus* Maleev, 1952 (*Pinacosaurus*)
*Szechuanosaurus* Young, 1942 (*nomen dubium*)

## T

| | | | |
|---|---|---|---|
| *Talarurus* Maleev, 1952 | Mongolia | Late Cretaceous | Ankylosauria: Ankylosauridae |
| *Tangvayosaurus* Allain and others, 1999 | Laos | Mid-Cretaceous | Sauropoda: Titanosauria |
| *Tanius* Wiman, 1929 | China | Late Cretaceous | Ornithopoda: Hadrosauridae |
| *Tanystrosuchus* Kuhn, 1963 (*Halticosaurus*) | | | |
| *Tarascosaurus* Le Loeuff & Buffetaut, 1991 | France | Late Cretaceous | Ornithopoda: Hadrosauridae |
| *Tarbosaurus* Maleev, 1955 | Mongolia | Late Cretaceous | Theropoda: Tyrannosauridae |
| *Tarchia* Maryanska, 1977 | Mongolia | Late Cretaceous | Ankylosauria: Ankylosauridae |
| *Tatisaurus* Simmons, 1965 | China | Early Jurassic | Thyreophora: family? |
| *Taveirosaurus* Telles-Antunes & Sigogneau-Russell, 1991 (*nomen dubium*) | | | |
| *Tawasaurus* Young, 1982 (*Lufengosaurus*) | | | |
| *Technosaurus* Chatterjee, 1984 | USA | Late Triassic | Ornithischia: family? |
| *Tecovasaurus* Hunt & Lucas, 1994 | USA | Late Triassic | Ornithischia: family? |
| *Tehuelchesaurus* Rich and others, 1999 | Argentina | Late Jurassic | Sauropoda: Eusauropoda |
| *Teinurosaurus* Nopcsa, 1928 (*Caudocoelus*) | | | |
| *Teishanosaurus* Dong, 1990 (*nomen nudum*) | | | |
| *Telmatosaurus* Nopcsa, 1903 | Romania | Late Cretaceous | Ornithopoda: Hadrosauridae |
| *Tenantosaurus* Brown, 1989 (*nomen nudum*) | | | |
| *Tendaguria* Bonaparte, Heinrich & Wild, 2000 | Tanzania | Late Jurassic | Sauropoda: Tendaguriidae |
| *Tenontosaurus* Ostrom, 1970 | USA | Early Cretaceous | Ornithopoda: Hypsilophodontidae |
| *Tetragonosaurus* Parks, 1931 (*Lambeosaurus*) | | | |
| *Texasetes* Coombs, 1995 | USA | Early Cretaceous | Ankylosauria: Ankylosauridae |
| *Teyuwasu* Kischlat, 1999 | Brazil | Late Triassic | Dinosauria: family? |
| *Thecocoelurus* Huene, 1923 (*Calamospondylus*) | | | |
| *Thecodontosaurus* Riley & Stutchbury, 1836 | England | Late Triassic | Prosauropoda: family? |
| *Thecospondylus* Seeley, 1882 (*nomen dubium*) | | | |
| *Therizinosaurus* Maleev, 1954 | Mongolia | Late Cretaceous | Theropoda: Therizinosauridae |
| *Therosaurus* Fitzinger, 1843 (*Iguanodon*) | | | |
| *Thescelosaurus* Gilmore, 1913 | Canada/USA | Late Cretaceous | Ornithopoda: family? |
| *Thespesius* Leidy, 1856 (*nomen dubium*) | | | |
| *Thotobolosaurus* Ellenberger, 1972 (*nomen nudum*) | | | |
| *Tianchisaurus* Dong, 1993 | China | Mid-Jurassic | Ankylosauria: Ankylosauridae |
| *Tianchungosaurus* Zhao, 1983 (*nomen nudum*) | | | |
| *Tianzhenosaurus* Pang & Cheng, 1998 | China | Late Cretaceous | Ankylosauria: Nodosauridae |
| *Tichosteus* Cope, 1877 (*nomen dubium*) | | | |
| *Tienshanosaurus* Young, 1937 | China | Late Jurassic | Sauropoda: Eusauropoda |
| *Timimus* Rich & Vickers-Rich, 1994 | Australia | Early Cretaceous | Theropoda: Coelurosauria |
| *Titanosaurus* Lydekker, 1877 | India | Late Cretaceous | Sauropoda: Titsanosauridae |
| *Titanosaurus* Marsh, 1877 (*Apatosaurus*) | | | |
| *Tochisaurus* Kurzanov & Osmólska, 1991 | Mongolia | Late Cretaceous | Theropoda; Troodontidae |
| *Tomodon* Leidy, 1865 (*Diplotomodon*) | | | |
| *Tonouchisaurus* Barsbold, 1994 (*nomen nudum*) | | | |
| *Tornieria* Sternfeld, 1911 (*Barosaurus*) | | | |
| *Torosaurus* Marsh, 1891 | USA | Late Cretaceous | Ceratopsia: Ceratopsidae |
| *Torvosaurus* Galton & Jensen, 1979 | USA | Late Jurassic | Theropoda: Torvosauridae |
| *Trachodon* Leidy, 1856 | USA | Late Cretaceous | Ornithopoda: Hadrosauridae |
| *Triceratops* Marsh, 1889 | USA | Late Cretaceous | Ceratopsia: Ceratopsidae |
| *Trimucrodon* Thulborn, 1973 (*nomen dubium*) | | | |
| *Troodon* Leidy, 1856 | USA | Late Cretaceous | Theropoda: Troodontidae |
| *Tsagantegia* Tumanova, 1993 | Mongolia | Late Cretaceous | Ankylosauria: Ankylosauridae |
| *Tsintaosaurus* Young, 1958 | China | Late Cretaceous | Ornithopoda: Hadrosauridae |
| *Tugulusaurus* Dong, 1973 (*nomen dubium*) | | | |
| *Tuojiangosaurus* Dong, Li, Zhou & Zhang, 1977 | China | Late Jurassic | Stegosauria: Stegosauridae |
| *Turanoceratops* Nesov & Kaznyshkina, 1989 | Kazakhstan | Late Cretaceous | Ceratopsia: Ceratopsidae |
| *Tylocephale* Maryanska & Osmólska, 1974 | Mongolia | Late Cretaceous | Pachycephalosauria: Pachycephalosauridae |
| *Tylosteus* Leidy, 1872 (*Pachycephalosaurus*) | | | |
| *Tyrannosaurus* Osborn, 1905 | USA/Canada | Late Cretaceous | Theropoda: Tyrannosauridae |
| *Tyreophorus* Huene, 1929 (*nomen nudum*) | | | |

## U

| | | | |
|---|---|---|---|
| *Udanoceratops* Kurzanov, 1992 | Mongolia | Late Cretaceous | Ceratopsia: Ceratopsidae |
| *Ugrosaurus* Cobabe & Fastovsky, 1987 (*Triceratops*) | | | |
| *Uintasaurus* Holland, 1919 (*Camarasaurus*) | | | |

*Ultrasauros* Jensen vide Olshevsky, 1991 (*Brachiosaurus*)
*Ultrasaurus* Kim, 1983 (*nomen dubium*)
*Ultrasaurus* Jensen, 1985 (*Ultrasauros*)
*Umarsaurus* Maryanska & Osmólska, 1981 (*nomen nudum*)

| | | | |
|---|---|---|---|
| *Unenlagia* Novas & Puerta, 1997 | Argentina | Late Cretaceous | Theropoda: Avialae |
| *Unquillosaurus* J. E. Powell, 1979 | Argentina | Late Cretaceous | Theropoda: Neotheropoda |
| *Utahraptor* Kirkland, Burge & Gaston, 1993 | USA | Early Cretaceous | Theropoda: Dromaseosauridae |

**V**

*Valdoraptor* Olshevsky, 1991 — England — Early Cretaceous — Theropoda: Carnosauria
*Valdosaurus* Galton, 1977 — England — Early Cretaceous — Ornithopoda: Dryosauridae?
*Variraptor* Le Loeuff & Buffetaut, 1998 — France — Late Cretaceous — Theropoda: Dromaeosauridae
*Vectensia* Delair, 1982 (*nomen nudum*)
*Vectisaurus* Hulke, 1879 (*Iguanodon*)
*Velocipes* Huene, 1932 (*nomen dubium*)
*Velociraptor* Osborn, 1924 — Mongolia — Late Cretaceous — Theropoda: Dromaeosauridae
*Velocisaurus* Bonaparte, 1991 — Argentina — Late Cretaceous — Theropoda: Abelisauroidea
*Volkheimeria* Bonaparte, 1979 — Argentina — Mid-Jurassic — Sauropoda: Eusauropoda
*Vulcanodon* Raath, 1972 — Zimbabwe — Early Jurassic — Sauropoda: Vulcanodontidae

**W**

*Wakinosaurus* Okazaki, 1992 (*nomen dubium*)
*Walgettosuchus* Huene, 1932 (*nomen dubium*)
*Walkeria* Chatterjee, 1987 (*Alwalkeria*)
*Walkersaurus* Welles, H. P. Powell & Pickering, 1995 (*nomen nudum*)
*Wannanosaurus* Hou, 1977 — China — Late Cretaceous — Pachycephalosauria: Pachycephalosauridae
*Wuerhosaurus* Dong, 1973 — China — Early Cretaceous — Stegosauria: Stegosauridae
*Wyomingraptor* Anonymous, 1997 (*nomen nudum*)

**X**

*Xenotarsosaurus* Martinez and others, 1986 — Argentina — Mid-Cretaceous — Theropoda: Abelisauridae
*Xiaosaurus* Dong & Tang, 1983 (*nomen dubium*)
*Xuanhanosaurus* Dong, 1984 — China — Mid-Jurassic — Theropoda: Tetanurae
*Xuanhuasaurus* Chao, 1985 (*nomen nudum*)

**Y**

*Yaleosaurus* Huene, 1932 (*Anchisaurus*)
*Yandusaurus* He, 1979 — China — Mid-Jurassic — Ornithopoda: family?
*Yangchuanosaurus* Dong, Chang, Li & Zhou, 1978 — China — Late Jurassic — Theropoda: Sinraptoridae
*Yaverlandia* Galton, 1971 — England — Early Cretaceous — Pachycephalosauria: Pachycephalosauridae
*Yimenosaurus* Bai, 1990 — China — Early Jurassic — Prosauropoda: family?
*Yingshanosaurus* Zhou, 1984 (*nomen nudum*)
*Yubasaurus* He, 1975 (*nomen nudum*)
*Yunnanosaurus* Young, 1942 — China — Early Jurassic — Prosauropoda: Yunnanosauridae

**Z**

*Zapsalis* Cope, 1876 (*Paronychodon*)
*Zephyrosaurus* Sues, 1980 — USA — Early Cretaceous — Ornithopoda: Zephyrosauridae
*Zigongosaurus* Hou, Chao & Chu, 1976 (*Mamenchisaurus*)
*Zizhongosaurus* Dong, Zhou & Zhang, 1983 — China — Early Cretaceous — Sauropoda: Vulcanodontidae?
*Zuniceratops* Wolfe & Kirkland, 1998 — USA — Late Cretaceous — Ceratopsia: Ceratopsidae

# BEST BOOKS

Benton, M. J. 1996. *The Penguin historical atlas of dinosaurs*. Penguin, London. Maps, locations, and up-to-date details about research on dinosaurs around the world.

Benton, M. J. 2000. *Vertebrate Palaeontology* (2nd edition). Blackwell Scientific, Oxford. A standard up-to-date text book.  ·

Currie, P. J. and Padian, K. 1997. *Encyclopedia of dinosaurs*. Academic Press, San Diego. A monster dictionary, good in parts.

Farlow, J. O. and Brett-Surman, M. K. (eds) 1998. *The complete dinosaur*. Indiana University Press, Bloomington. The best compilation of articles by leading experts on the life and times of the dinosaurs.

Fastovsky, D. E. and Weishampel, D. B. 1996. *The evolution and extinction of the dinosaurs*. Cambridge University Press. The best testbook devoted to dinosaurs.

Haines, T. 1999. *Walking with dinosaurs*. BBC, London. The book of the series.

Norman, D. B. 1985. *The illustrated encyclopedia of dinosaurs*. Salamander, London. The classic dinosaur book, with beautiful artwork by John Sibbick.

Weishampel, D. B., Dodson, P, and Osmólska, H. (eds) 1990. *The Dinosauria*. University of California Press, Berkeley. This is the 'bible' of dinosaur studies. A second edition is due in 2001.

# BEST WEB SITES

There are thousands of sites of the World Wide Web that deal with dinosaurs.
Most of these do not offer much information. Here are some of the best.

BBC Official 'Walking with Dinosaurs' site
**http://www.bbc.co.uk/dinosaurs**
Dinobase: a complete listing of dinosaurs and basic data
**http://palaeo.gly.bristol.ac.uk/dinobase/dinopage.html**
Background information about Walking with Dinosaurs
**http://palaeo.gly.bris.ac.uk/walking.html**
Dinosaur Society (UK)
**http://www.hmag.gla.ac.uk/dinosoc/**
Paleonet, a server for all palaeontologists
**http://www.ucmp.berkeley.edu/Paleonet/** (for North America)
**http://www.nhm.ac.uk/paleonet/** (for users elsewhere)
Society of Vertebrate Paleontology
**http://alnus.uel.ac.uk/svp/**
American Museum of Natural History
**http://www.amnh.org/**
Carnegie Museum of Natural History, Pittsburg
**http://www.clpgh.org/cmnh/**
Field Museum of Natural History, Chicago
**http://www.uic.edu/orgs/paleo/homepage.html**
Museum of Paleontology, Berkeley, California
**http://www.ucmp.berkeley.edu/**
Natural History Museum, London
**http://www.nhm.ac.uk/**
New Mexico Museum of Natural History, Albuquerque
**http://www.nmmnh-abq.mus.nm.us/nsmnh/**
Royal Tyrrell Museum, Drumheller, Canada
**http://tyrrell.magtech.ab.ca/home.html**
Smithsonian Institution, Washington, D.C.
**http://nmnh.si.edu/departments/paleo.html**

# INDEX